APICULTURE
Basic Fundamentals

APICULTURE
Basic Fundamentals

by
Paramjit Bhasin | Abhijeet Acharya

Wordpen Academics
Publishing Science That Matter

ISBN: 978-93-6084-195-9 (Hardback)

Published by : **WORDPEN ACADEMICS**
4736/23, Ansari Road, Darya Ganj,
New Delhi – 110 002
Phone: +91-011-35004336
E-mail: info@wordpenacademics.com

Printed at **:** **Replika Press Pvt. Ltd.**

Exclusive Distributor: ***Bio-Green Books, New Delhi***

Preface

"Apiculture: Basic Fundamentals" is an invaluable resource that serves as a gateway to the fascinating world of beekeeping and honey production. This book is a testament to the enduring significance of apiculture as a vital agricultural practice, offering readers a comprehensive understanding of the fundamental principles, techniques, and best practices essential for success in beekeeping endeavors.

The art and science of apiculture have been intertwined with human civilization for millennia, with bees playing a pivotal role not only as producers of honey but also as crucial pollinators essential for the health and productivity of ecosystems and agricultural crops. Against the backdrop of growing environmental concerns and the imperative of sustainable food production, the relevance of beekeeping as a livelihood and conservation tool has never been more pronounced.

As editors of this volume, we have endeavored to compile a wealth of knowledge from seasoned beekeepers, researchers, and industry experts, each contributing their expertise to illuminate every aspect of beekeeping – from hive management and bee biology to honey extraction and marketing strategies. Whether you are a novice enthusiast or a seasoned professional, this book offers practical guidance, insightful tips, and scientific insights to enhance your understanding and proficiency in apiculture.

Our overarching goal in presenting this book is to inspire a new generation of beekeepers and foster a deeper appreciation for the vital role of bees in sustaining biodiversity and food security. By promoting responsible beekeeping practices and raising awareness about the importance of pollinator conservation, we hope to empower individuals and communities to engage in the stewardship of bees and their habitats.

We extend our sincere gratitude to all the contributors who have generously shared their knowledge, experiences, and passion for apiculture, as well as to the readers who embark on this enriching journey with us. May this volume serve as a springboard for exploration, learning, and discovery, paving the way for a brighter future where humans and bees thrive in harmony.

Paramjit Bhasin | Abhijeet Acharya

Contents

1

Apiculture: An Introduction

Beekeeping is the maintenance of honey bee colonies, commonly in hives, by humans. A beekeeper keeps bees in order to collect honey and other products of the hive, to pollinate crops, or to produce bees for sale to other beekeepers. A location where bees are kept is called an apiary or "bee yard".

There are more than 20,000 species of wild bees. Many species are solitary and many others rear their young in burrows and small colonies, (e.g. bumblebees). Beekeeping, or apiculture, is concerned with the practical management of the social species of honey bees, which live in large colonies of up to 100,000 individuals. In Europe and America the species universally managed by beekeepers is the Western honey bee (Apis mellifera). This species has several sub-species or regional varieties, such as the Italian bee , European dark bee, and the Carniolan honey bee. In the tropics, other species of social bee are managed for honey production, including Apis cerana.

All of the Apis mellifera sub-species are capable of inter-breeding and hybridizing. Many bee breeding companies strive to selectively breed and hybridize varieties to produce desirable qualities: disease and parasite resistance, good honey production, swarming behaviour reduction, prolific breeding, and mild disposition.

Some of these hybrids are marketed under specific brand names, such as the Buckfast Bee or Midnite Bee. The advantages of the initial F1 hybrids produced by these crosses include: hybrid vigor, increased honey productivity, and greater disease resistance. The disadvantage is that in subsequent generations these advantages may fade away and hybrids tend to be very defensive and aggressive.

STUDY OF HONEY BEES

It was not until the 18th century that European natural philosophers undertook the scientific study of bee colonies and began to understand the complex and hidden world of bee biology. Preeminent among these scientific pioneers were Swammerdam, René Antoine Ferchault de Réaumur, Charles Bonnet, and the blind Swiss scientist Francois Huber. Swammerdam and Réaumur were among the first to use a microscope and dissection to understand the internal biology of honey bees. Réaumur was among the first to construct a glass walled observation hive to better observe activities within hives.

He observed queens laying eggs in open cells, but still had no idea of how a queen was fertilized; nobody had ever witnessed the mating of a queen and drone and many theories held that queens were "self-fertile," while others believed that a vapor or "miasma" emanating from the drones fertilized queens without direct physical contact. Huber was the first to prove by observation and experiment that queens are physically inseminated by drones outside the confines of hives, usually a great distance away.

Following Réaumur's design, Huber built improved glass-walled observation hives and sectional hives which could be opened, like the leaves of a book, to inspect individual wax combs; this greatly improved the direct observation of activity within a hive. Although he became blind before he was twenty, Huber employed a secretary, Francois Burnens, to make daily observations, conduct careful experiments, and to keep accurate notes over a period of more than twenty years.

Huber confirmed that a hive consists of one queen who is the mother of all the female workers and male drones in the colony. He was also the first to confirm that mating with drones takes place outside of hives and that queens are inseminated by a number of successive matings with male drones, high in the air at a great distance from their hive. Together, he and Burnens dissected bees under the microscope and were among the first to describe the ovaries and spermatheca, or sperm store, of queens as well as the penis of male drones.

Invention of the movable comb hive

Early forms of honey collecting entailed the destruction of the entire colony when the honey was harvested. The wild hive was crudely broken into, using smoke to suppress the bees, the honeycombs were torn out and smashed up — along with the eggs, larvae and honey they contained. The liquid honey from the destroyed brood nest was crudely strained through a sieve or basket. This was destructive and unhygienic, but for hunter-gatherer societies this did not matter, since the honey was generally consumed immediately and there were always more wild colonies to exploit. But in settled societies the destruction of the bee colony meant the loss of a valuable resource; this drawback made beekeeping both inefficient and something of a "stop and start" activity.

There could be no continuity of production and no possibility of selective breeding, since each bee colony was destroyed at harvest time, along with its precious queen. During the medieval period abbeys and monasteries were centers of beekeeping, since beeswax was highly prized for candles and fermented honey was used to make alcoholic mead in areas of Europe where vines would not grow.

The 18th and 19th centuries saw successive stages of a revolution in beekeeping, which allowed the bees themselves to be preserved when taking the harvest.

Intermediate stages in the transition from the old beekeeping to the new were recorded for example by Thomas Wildman in 1768/1770, who described advances over the destructive old skep-based beekeeping so that the bees no longer had to be killed to harvest the honey. Wildman for example fixed a parallel array of wooden bars across the top of a straw hive or skep (with a separate straw top to be fixed on later) "so that there are in all seven bars of deal" [in a 10-inch-diameter (250 mm) hive] "to which the bees fix their combs". He also described using such hives in a multi-storey configuration, foreshadowing the modern use of supers: he described adding (at a proper time) successive straw hives below, and eventually removing the ones above when free of brood and filled with honey, so that the bees could be separately preserved at the harvest for a following season.

Wildman also described a further development, using hives with "sliding frames" for the bees to build their comb, foreshadowing more modern uses of movable-comb hives. Wildman's book acknowledged the advances in knowledge of bees previously made by Swammerdam, Maraldi, and de Réaumur — he included a lengthy translation of Réaumur's account of the natural history of bees — and he also described the initiatives of others in designing hives for the preservation of bee-life when taking the harvest, citing in particular reports from Brittany dating from the 1750s, due to Comte de la Bourdonnaye.

The 19th Century saw this revolution in beekeeping practice completed through the perfection of the movable comb hive by Lorenzo Lorraine Langstroth, a descendant of Yorkshire farmers who emigrated to the United States. Langstroth was the first person to make practical use of Huber's earlier discovery that there was a specific spatial measurement between the wax combs, later called "the bee space", which bees would not block with wax, but kept as a free passage. Having determined this "bee space" (between 5 and 8 mm, or 1/4 to 3/8"), Langstroth then designed a series of wooden frames within a rectangular hive box, carefully maintaining the correct space between successive frames, and found that the bees would build parallel honeycombs in the box without bonding them to each other or to the hive walls. This enables the beekeeper to slide any frame out of the hive for inspection, without harming the bees or the comb, protecting the eggs, larvae and pupae contained within the cells. It also meant that combs containing honey could be gently removed and the honey extracted without destroying the comb.

Domestication of wild bees

At some point humans began to attempt to domesticate wild bees in artificial hives made from hollow logs, wooden boxes, pottery vessels, and woven straw baskets or "skeps". Honeybees were kept in Egypt from antiquity. On the walls of the sun temple of Nyuserre Ini from the 5th Dynasty, before 2422 BCE, workers are depicted blowing smoke into hives as they are removing honeycombs. Inscriptions detailing the production of honey are found on the tomb of Pabasa from the 26th Dynasty (circa 650 BCE), depicting pouring honey in jars and cylindrical hives. Sealed pots of honey were found in the grave goods of Pharaohs such as Tutankhamun.

In prehistoric Greece (Crete and Mycenae), there existed a system of high-status apiculture, as can be concluded from the finds of hives, smoking pots, honey extractors and other beekeeping paraphernalia in Knossos. Beekeeping was considered a highly valued industry controlled by beekeeping overseers — owners of gold rings depicting apiculture scenes rather than religious ones as they have been reinterpreted recently, contra Sir Arthur Evans.

Archaeological finds relating to beekeeping have been discovered at Rehov, a Bronze and Iron Age archaeological site in the Jordan Valley, Israel. Thirty intact hives, made of straw and unbaked clay, were discovered by archaeologist Amihai Mazar of the Hebrew University of

Jerusalem in the ruins of the city, dating from about 900 BCE. The hives were found in orderly rows, three high, in a manner that could have accommodated around 100 hives, held more than 1 million bees and had a potential annual yield of 500 kilograms of honey and 70 kilograms of beeswax, according to Mazar, and are evidence that an advanced honey industry existed in ancient Israel 3,000 years ago. Ezra Marcus, an expert from the University of Haifa, said the finding was a glimpse of ancient beekeeping seen in texts and ancient art from the Near East.

The art of beekeeping appeared in ancient China for a long time and hardly traceable to its origin. In the book "Golden Rules of Business Success" written by Fan Li (or Tao Zhu Gong) during the Spring and Autumn Period there are some parts mentioning the art of beekeeping and the importance of the quality of the wooden box for bee keeping that can affect the quality of its honey.

Wild honey harvesting

Collecting honey from wild bee colonies is one of the most ancient human activities and is still practiced by aboriginal societies in parts of Africa, Asia, Australia, and South America. Some of the earliest evidence of gathering honey from wild colonies is from rock paintings, dating to around 13,000 BCE. Gathering honey from wild bee colonies is usually done by subduing the bees with smoke and breaking open the tree or rocks where the colony is located, often resulting in the physical destruction of the nest location.

Evolution of hive designs

Langstroth's design for movable comb hives was seized upon by apiarists and inventors on both sides of the Atlantic and a wide range of moveable comb hives were designed and perfected in England, France, Germany and the United States. Classic designs evolved in each country: Dadant hives and Langstroth hives are still dominant in the USA; in France the De-Layens trough-hive became popular and in the UK a British National Hive became standard as late as the 1930s although in Scotland the smaller Smith hive is still popular.

In some Scandinavian countries and in Russia the traditional trough hive persisted until late in the 20th Century and is still kept in some areas. However, the Langstroth and Dadant designs remain ubiquitous in the USA and also in many parts of Europe, though Sweden, Denmark, Germany, France and Italy all have their own national hive designs.

Regional variations of hive evolved to reflect the climate, floral productivity and the reproductive characteristics of the various subspecies of native honey bee in each bio-region.

The differences in hive dimensions are insignificant in comparison to the common factors in all these hives: they are all square or rectangular; they all use movable wooden frames; they all consist of a floor, brood-box, honey-super, crown-board and roof. Hives have traditionally been constructed of cedar, pine, or cypress wood, but in recent years hives made from injection molded dense polystyrene have become increasingly important.

Hives also use queen excluders between the brood-box and honey supers to keep the queen from laying eggs in cells next to those containing honey intended for consumption. Also, with the advent in the 20th century of mite pests, hive floors are often replaced for part of (or the whole) year with a wire mesh and removable tray.

Pioneers of practical and commercial beekeeping

The 19th Century produced an explosion of innovators and inventors who perfected the design and production of beehives, systems of management and husbandry, stock improvement by selective breeding, honey extraction and marketing. Preeminent among these innovators were:

Jan Dzier¿on, was the father of modern apiology and apiculture. All modern beehives are descendants of his design.

L. L. Langstroth, revered as the "father of American apiculture", no other individual has influenced modern beekeeping practice more than Lorenzo Lorraine Langstroth. His classic book The Hive and Honey-bee was published in 1853.

Moses Quinby, often termed 'the father of commercial beekeeping in the United States', author of Mysteries of Bee-Keeping Explained.

Amos Root, author of the A B C of Bee Culture which has been continuously revised and remains in print to this day. Root pioneered the manufacture of hives and the distribution of bee-packages in the United States.

Dr. C.C. Miller was one of the first entrepreneurs to actually make a living from apiculture. By 1878 he made beekeeping his sole business activity. His book, Fifty Years Among the Bees, remains a classic and his influence on bee management persists to this day.

Major Francesco De Hruschka was an Italian military officer who made one crucial invention that catalyzed the commercial honey industry. In 1865 he invented a simple machine for extracting honey from the comb by means of centrifugal force. His original idea was simply to support combs in a metal framework and then spin them around within a container to collect honey as it was thrown out by centrifugal force. This meant that honeycombs could be returned to a hive undamaged but empty, saving the bees a vast amount of work, time, and materials. This single invention greatly improved the efficiency of honey harvesting and catalysed the modern honey industry.

In the U.K. practical beekeeping was led in the early 20th century by a few men, pre-eminently Brother Adam and his Buckfast bee and R.O.B. Manley, author of many titles, including 'Honey Production In The British Isles' and inventor of the Manley frame, still universally popular in the U.K. Other notable British pioneers include William Herrod-Hempsall and Gale.

In India, N. K. Garg is known as the pioneer of starting apiculture. He is the one who helped people in taking apiculture commercially. He started it in early 1990's.

Traditional beekeeping

Fixed comb hives

A fixed comb hive is a hive in which the combs cannot be removed or manipulated for management or harvesting without permanently damaging the comb. Almost any hollow structure can be used for this purpose, such as a log gum, skep or a clay pot. Fixed comb hives are no longer in common use in most places, and are illegal in some places that require inspection for problems such as varroa and American foulbrood.

Modern beekeeping

Movable frame hives

In the United States, the Langstroth hive is commonly used. The Langstroth was the first successful top-opened hive with movable frames, and other designs of hive have been based on it. Langstroth hive was however a descendant of Jan Dzierzon's Polish hive designs. In the United Kingdom, the most common type of hive is the British

National Hive, which can hold Hoffman, British Standard or popular Manley frames, but it is not unusual to see some other sorts of hive (Smith, Commercial and WBC, rarely Langstroth). Straw skeps, bee gums, and unframed box hives are now unlawful in most US states, as the comb and brood cannot be inspected for diseases. However, straw skeps are still used for collecting swarms by hobbyists in the UK, before moving them into standard hives.

Top-bar hives

A growing number of amateur beekeepers are adopting various top-bar hives similar to the type commonly found in Africa. Top bar hives were originally used as traditional beekeeping a method in both Greece and Vietnam. These have no frames and the honey-filled comb is not returned to the hive after extraction, as it is in the Langstroth hive. Because of this, the production of honey is likely to be somewhat less than that of a Langstroth hive. Top bar hives are mostly kept by people who are more interested in having bees in their garden than in honey production per se.

Some of the most well known top-bar hives are the Kenyan Top Bar Hive (KTBH) with sloping sides, the Tanzanian Top Bar Hive, which has straight sides and the Vertical Top Bar Hives such as the Warre or "People's Hive" designed by Abbe Warre in the mid 1900's.

The initial costs and equipment requirements are far lower. Often scrap wood or #2 or #3 pine is able to be used with a nice hive as the outcome. Top-bar hives also offer some advantages in interacting with the bees and the amount of weight that must be lifted is greatly reduced. Top-bar Hives are being widely used in developing countries in Africa and Asia as a result of the 'Bees For Development' program. There is a growing number of beekeepers in the U.S. using various top-bar hives within the past 2 years or so.

Smoker

Smoke is the beekeeper's third line of defense. Most beekeepers use a "smoker" — a device designed to generate smoke from the incomplete combustion of various fuels. Smoke calms bees; it initiates a feeding response in anticipation of possible hive abandonment due to fire. Smoke also masks alarm pheromones released by guard bees or when bees are squashed in an inspection. The ensuing confusion creates an opportunity for the beekeeper to open the hive and work without triggering a defensive reaction. In addition, when a bee consumes honey the bee's abdomen distends, supposedly making it difficult to make the necessary flexes to sting, though this has not been tested scientifically.

Smoke is of questionable use with a swarm, because swarms do not have honey stores to feed on in response. Usually smoke is not needed, since swarms tend to be less defensive, as they have no stores to defend, and a fresh swarm will have fed well from the hive.

Many types of fuel can be used in a smoker as long as it is natural and not contaminated with harmful substances. These fuels include hessian, twine, burlap, pine needles, corrugated cardboard, and mostly rotten or punky wood. Indian beekeepers, especially in Kerala, often use coconut fibers as they are readily available, safe, and of negligible expense. Some beekeeping supply sources also sell commercial fuels like pulped paper and compressed cotton, or even aerosol cans of smoke.

Some beekeepers are using "liquid smoke" as a safer, more convenient, alternative. It is a water-based solution that is sprayed onto the bees from a plastic spray bottle.

PROTECTIVE CLOTHING

While knowledge of the bees is the first line of defense, most beekeepers also wear some protective clothing. Novice beekeepers usually wear gloves and a hooded suit or hat and veil. Experienced beekeepers sometimes elect not to use gloves because they inhibit delicate manipulations. The face and neck are the most important areas to protect, so most beekeepers will at least wear a veil.

Defensive bees are attracted to the breath, and a sting on the face can lead to much more pain and swelling than a sting elsewhere, while a sting on a bare hand can usually be quickly removed by fingernail scrape to reduce the amount of venom injected.

The protective clothing is generally light colored (but not colorful) and of a smooth material. This provides the maximum differentiation from the colony's natural predators (bears, skunks, etc.) which tend to be dark-colored and furry.

The 'stings' retained in the fabric of the clothing will continue to pump out an alarm pheromone that actually attracts aggressive action and further stinging attacks. Washing suits regularly, and rinsing gloved hands in vinegar will minimize attraction.

Effects of stings and of protective measures

Some beekeepers believe that the more stings a beekeeper receives, the less irritation each causes, and they consider it important for safety of the beekeeper to be stung a few times a season. Beekeepers have high levels of antibodies reacting to the major antigen of bee venom, phospholipase A2 (PLA). Antibodies correlate with the frequency of bee stings.

The entry of venom into the body from bee-stings may also be hindered and reduced by protective clothing which allows the wearer to remove stings and venom sacs simply with a tug on the clothing.

Natural beekeeping

There is a current movement that eschews chemicals in beekeeping and believes that health issues in bees can most effectively be addressed by reversing trends that disrespect the needs of the bees themselves. Crop spraying, unnatural conditions in which bees are moved thousands of miles to pollinate commercial crops, frequent opening of the hive for inspection, artificial insemination of queens, routine medication and sugar water feeding are all thought to contribute to a general weakening of the constitution of the honey bee.

Practitioners of 'natural beekeeping' tend to use variations of the top-bar hive, which is a simple design that retains the concept of movable comb without the use of frames or foundation. The horizontal top-bar hive, as championed by Marty Hardison, Michael Bush, Philip Chandler, Dennis Murrell and others, can be seen as a modernization of hollow log hives, with the addition of wooden bars of specific width from which bees hang their combs. Its widespread adoption in recent years can be attributed to the publication in 2007 of The Barefoot Beekeeper by Philip Chandler, which challenged many aspects of modern beekeeping and offered the horizontal top-bar hive as a viable alternative to the ubiquitous Langstroth-syle movable-frame hive.

The most popular vertical top-bar hive is probably the Warré hive, based on a design by the French priest Abbé Émile Warré (1867-1951) and popularized by Dr. David Heaf in his English translation of Warré's book L'Apiculture por Tous as Beekeeping For All.

Natural beekeeping is characterized by a willingness to hand most of the control to the bees themselves, and to minimize interference in their lives. Practitioners expect to take honey only when the bees' needs have first been taken care of, and the feeding of sugar is discouraged except as an emergency measure.

Bee colonies

Castes

A colony of bees consists of three castes of bee:

- A queen bee, which is normally the only breeding female in the colony;
- A large number of female worker bees, typically 30,000–50,000 in number;
- A number of male drones, ranging from thousands in a strong hive in spring to very few during dearth or cold season.

The queen is the only sexually mature female in the hive and all of the female worker bees and male drones are her offspring. The queen may live for up to three years or more and may be capable of laying half a million eggs or more in her lifetime. At the peak of the breeding season, late spring to summer, a good queen may be capable of laying 3,000 eggs in one day, more than her own body weight. This would be exceptional however; a prolific queen might peak at 2,000 eggs a day, but a more average queen might lay just 1,500 eggs per day.

The queen is raised from a normal worker egg, but is fed a larger amount of royal jelly than a normal worker bee, resulting in a radically different growth and metamorphosis. The queen influences the colony by the production and dissemination of a variety of pheromones or "queen substances". One of these chemicals suppresses the development of ovaries in all the female worker bees in the hive and prevents them from laying eggs.

MATING OF QUEENS

The queen emerges from her cell after 15 days of development and she remains in the hive for 3–7 days before venturing out on a mating flight. Mating flight is otherwise known as 'nuptial flight'. Her first orientation flight may only last a few seconds, just enough to mark the position of the hive. Subsequent mating flights may last from 5 minutes to 30 minutes, and she may mate with a number of male drones on each flight. Over several matings, possibly a dozen or more, the queen will receive and store enough sperm from a succession of drones to fertilize hundreds of thousands of eggs. If she does not manage to leave the hive to mate — possibly due to bad weather or being trapped within part of the hive — she will remain infertile and become a 'drone layer', incapable of producing female worker bees. Worker bees will sometimes kill a non-performing queen, and produce another. Without a properly performing queen, the hive is doomed.

Mating takes place at some distance from the hive and often several hundred feet up in the air; it is thought that this separates the strongest drones from the weaker ones - ensuring that only the fastest and strongest drones get to pass on their genes.

Female worker bees

Almost all the bees in a hive are female worker bees. At the height of summer when activity in the hive is frantic and work goes on non-stop, the life of a worker bee may be as short as 6 weeks; in late autumn, when no brood is being raised and no nectar is being harvested, a young bee may live for 16 weeks, right through the winter. During its life a worker bee performs different work functions in the hive which are largely dictated by the age of the bee.

Period	Work activity
Days 1-3	Cleaning cells and incubation
Day 3-6	Feeding older larvae
Day 6-10	Feeding younger larvae
Day 8-16	Receiving honey and pollen from field bees
Day 12-18	Wax making and cell building
Day 14 onwards	Entrance guards; nectar and pollen foraging

Male bees (drones)

Drones are the largest bees in the hive (except for the queen), at almost twice the size of a worker bee. They do no work, do not forage for pollen or nectar and are only produced in order to mate with new queens and fertilize them on their mating flights. A bee colony will generally start to raise drones a few weeks before building queen cells in order to supersede a failing queen or in preparation for swarming. When queen raising for the season is over, the bees in colder climates will drive the drones out of the hive to die, biting and tearing their legs and wings.

Differing stages of development

Stage of development	Queen	Worker	Drone
Egg	3 days	3 days	3 days
Larva	8 days	10 days	13 days
Pupa	4 days	8 days	8 days
Total	15 days	21 days	24 days

Structure of a bee colony

A domesticated bee colony is normally housed in a rectangular hive body, within which eight to ten parallel frames house the vertical plates of honeycomb which contain the eggs, larvae, pupae and food for the colony.

If one were to cut a vertical cross-section through the hive from side to side, the brood nest would appear as a roughly ovoid ball spanning 5-8 frames of comb. The two outside combs at each side of the hive tend to be exclusively used for long-term storage of honey and pollen.

Within the central brood nest, a single frame of comb will typically have a central disk of eggs, larvae and sealed brood cells which may extend almost to the edges of the frame. Immediately above the brood patch an arch of pollen-filled cells extends from side to side, and above that again a broader arch of honey-filled cells extends to the frame tops.

The pollen is protein-rich food for developing larvae, while honey is also food but largely energy rich rather than protein rich. The nurse bees which care for the developing brood secrete a special food called 'royal jelly' after feeding themselves on honey and pollen. The amount of royal jelly which is fed to a larva determines whether it will develop into a worker bee or a queen.

Apart from the honey stored within the central brood frames, the bees store surplus honey in combs above the brood nest. In modern hives the beekeeper places separate boxes, called 'supers', above the brood box, in which a series of shallower combs is provided for storage of honey. This enables the beekeeper to remove some of the supers in the late summer, and to extract the surplus honey harvest, without damaging the colony of bees and its brood nest below. If all the honey is 'stolen', including the amount of honey needed to survive winter, the beekeeper must replace these stores by feeding the bees sugar or corn syrup in autumn.

Annual cycle of a bee colony

The development of a bee colony follows an annual cycle of growth which begins in spring with a rapid expansion of the brood nest, as soon as pollen is available for feeding larvae. Some production of brood may begin as early as January, even in a cold winter, but breeding accelerates towards a peak in May (in the northern hemisphere), producing an abundance of harvesting bees synchronized to the main 'nectar flow' in that region.

Each race of bees times this build-up slightly differently, depending on how the flora of its original region blooms. Some regions of Europe have two nectar flows: one in late spring and another in late August. Other regions have only a single nectar flow. The skill of the beekeeper lies in predicting when the nectar flow will occur in his area and in trying to ensure that his colonies achieve a maximum population of harvesters at exactly the right time.

The key factor in this is the prevention or skillful management of the swarming impulse. If a colony swarms unexpectedly and the beekeeper does not manage to capture the resulting swarm, he is likely to harvest significantly less honey from that hive, since he will have lost half his worker bees at a single stroke. If, however, he can use the swarming impulse to breed a new queen but keep all the bees in the colony together, he will maximize his chances of a good harvest.

It takes many years of learning and experience to be able to manage all these aspects successfully, though owing to variable circumstances many beginners will often achieve a good honey harvest.

Formation of new colonies

Colony reproduction: swarming and supersedure

All colonies are totally dependent on their queen, who is the only egg-layer. However, even the best queens live only a few years and one or two years longevity is the norm. She can choose whether or not to fertilize an egg as she lays it; if she does so, it develops into a female worker bee; if she lays an unfertilized egg it becomes a male drone.

She decides which type of egg to lay depending on the size of the open brood cell which she encounters on the comb; in a small worker cell she lays a fertilized egg; if she finds a much larger drone cell she lays an unfertilized drone egg.

All the time that the queen is fertile and laying eggs she produces a variety of pheromones which control the behavior of the bees in the hive; these are commonly called 'queen substance' but in reality there are various different pheromones with different functions. As the queen ages she begins to run out of stored sperm and her pheromones begin to fail. At some point, inevitably, the queen begins to falter and the bees will decide to replace her by creating a new queen from one of her worker eggs. They may do this because she has been damaged (lost a leg or an antenna), because she has run out of sperm and cannot lay fertilized eggs (has become a 'drone laying queen') or because her pheromones have dwindled to a point where they cannot control all the bees in the hive anymore.

At this juncture the bees will produce one or more queen cells by modifying existing worker cells which contain a normal female egg. However, there are two distinct behaviors which the bees pursue:

1. Supersedure: queen replacement within one hive without swarming
2. Swarm cell production: the division of the hive into two colonies by swarming

Different sub-species of Apis mellifera exhibit differing swarming characteristics which reflect their evolution in different ecotopes of the European continent. In general the more northerly black races are said to swarm less and supersede more, whereas the more southerly yellow and grey varieties are said to swarm more frequently. The truth is complicated because of the prevalence of cross-breeding and hybridization of the sub species and opinions differ.

Supersedure is highly valued as a behavioral trait by beekeepers because a hive that supersedes its old queen does not swarm and so no stock is lost; it merely creates a new queen and allows the old one to fade away, or alternatively she is killed when the new queen emerges. When superseding a queen the bees will produce just one or two queen cells, characteristically in the center of the face of a broodcomb.

In swarming, by contrast, a great many queen cells are created — typically a dozen or more — and these are located around the edges of a broodcomb, most often at the sides and the bottom.

Once either process has begun, the old queen will normally leave the hive with the hatching of the first queen cells. When she leaves the hive the old queen is accompanied by a large number of bees, predominantly young bees (wax-secreters), who will form the basis of the new hive. Scouts are sent out from the swarm to find suitable hollow trees or rock crevices and as soon as one is found the entire swarm moves in, building new wax brood combs within a matter of hours using the honey stores which the young bees have filled themselves with before leaving the old hive. Only young bees can secrete wax from special abdominal segments and this is why there tend to be more young bees than old in swarms

FACTORS THAT TRIGGER SWARMING

It is generally accepted that a colony of bees will not swarm until they have completed all of their brood combs, i.e. filled all available space with eggs, larvae and brood. This generally occurs in late Spring at a time when the other areas of the hive are rapidly filling with honey stores. So one key trigger of the swarming instinct is when the queen has no more room to lay eggs and the hive population is becoming very congested. Under these conditions a prime swarm may issue with the queen, resulting in a halving of the population within the hive, leaving the old colony with a large number of hatching bees. The queen who leaves finds herself in a new hive with no eggs and no larvae but lots of energetic young bees who create a new set of brood combs from scratch in a very short time.

Another important factor in swarming is the age of the queen. Those under a year in age are unlikely to swarm unless they are extremely crowded, while older queens have swarming predisposition.

Beekeepers monitor their colonies carefully in spring and watch for the appearance of queen cells, which are a dramatic signal that the colony is determined to swarm.

When a colony has decided to swarm, queen cells are produced in numbers varying to a dozen or more. When the first of these queen cells is sealed, after eight days of larval feeding, a virgin queen will pupate and be due to emerge seven days after sealing. Before leaving, the worker bees fill their stomachs with honey in preparation for the creation of new honeycombs in a new home. This cargo of honey also makes swarming bees less inclined to sting, and a newly issued swarm is noticeably gentle for up to 24 hours — often capable of being handled without gloves or veil by a beekeeper.

This swarm is looking for shelter. A beekeeper may capture it and introduce it into a new hive, helping to meet this need. Otherwise, it will return to a feral state, in which case it will find shelter in a hollow tree, an excavation, an abandoned chimney, or even behind shutters.

Back at the original hive, the first virgin queen to emerge from her cell will immediately seek to kill all her rival queens who are still waiting to emerge from their cells. However, usually the bees deliberately prevent her from doing this, in which case, she too will lead a second swarm from the hive. Successive swarms are called 'after-swarms' or 'casts' and can be very small, often with just a thousand or so bees, as opposed to a prime swarm which may contain as many as ten to twenty thousand bees.

Small after-swarms have less chance of survival and, depleting the original hive, may threaten its survival as well. When a hive has swarmed despite the beekeeper's preventative efforts, a good management practice is to give the depleted hive a couple frames of open brood with eggs. This helps replenish the hive more quickly, and gives a second opportunity to raise a queen, if there is a mating failure.

Each race or sub-species of honey bee has its own swarming characteristics. Italian bees are very prolific and inclined to swarm; Northern European black bees have a strong tendency to supersede their old queen, without swarming. These differences are the result of differing evolutionary pressures in the regions where each sub-species evolved.

Artificial swarming

When a colony accidentally loses its queen, it is said to be 'queenless'. The workers realize that the queen is absent after as little as an hour, as her pheromones fade in the hive. The colony cannot survive without a fertile queen laying eggs to renew the population. So the workers select cells containing eggs aged less than three days and enlarge these cells dramatically to form 'emergency queen cells'. These appear similar to large peanut-like structure about an inch long, which hangs from the center or side of the brood combs.

The developing larva in a queen cell is fed differently from an ordinary worker-bee, receiving in addition to the normal honey and pollen a great deal of royal jelly, a special food secreted by young 'nurse bees' from the hypopharyngeal gland. This special food dramatically alters the growth and development of the larva so that, after metamorphosis and pupation, it emerges from the cell as a queen bee. The queen is the only bee in a colony which has fully developed ovaries and she secretes a pheromone which suppresses the normal development of ovaries in all her workers.

Beekeepers use the ability of the bees to produce new queens in order to increase their colonies, a procedure called splitting a colony. In order to do this, they remove several brood combs from a healthy hive, taking care that the old queen is left behind. These combs must contain eggs or larvae less than three days old which will be covered by young 'nurse bees' which care for the brood and keep it warm. These brood combs and attendant nurse bees are then placed into a small 'nucleus hive' along with other combs containing honey and pollen.

Urban or backyard beekeeping

Related to natural beekeeping, urban beekeeping is an attempt to revert to a less industrialized way of obtaining honey by utilizing small-scale colonies that pollinate urban gardens. Urban apiculture has undergone a renaissance in the 2000s. Paris, Berlin, London, Tokyo and Washington, D.C., are among beekeeping cities. Until 2010, beekeeping was banned in New York City and punishable with a $2,000 fine. Urban beekeeping is commonly practiced in areas that have a pesticide ban.

Bee-Keeping

In order to maximise agricultural production, honeybee can be used as an important input. About 85 per cent crop plants are cross-pollinated, as they need to receive pollen from other plants of the same species with the help of external agents. One of the most important external agents is the honeybee. A few colonies of honeybees are placed in the field when the crop is in flowering stage.

When pressed in to service they would make several thousand forages for pollination. The abundance of pollinators helps in early setting of seeds resulting in early and more uniform crop yield. Honeybees also produce honey, bee wax and royal jelly thus giving additional benefits to the farmers. About 50 million ha. of land in the country are under bee dependent crops like fruit, vegetables, oilseeds, legumes and pulses. Many of these crops require three to nine bee colonies per ha.

The National Bee Board which is looking after the developmental activities of bee keeping has been recently reconstituted with Secretary (A&C) as Chairman and Horticulture Commissioner as Member Secretary.

Beekeeping In India

Honey and beekeeping have a long history in India. Honey was the first sweet food tasted by the ancient Indian inhabiting rock shelters and forests. He hunted bee hives for this gift of god. India has some of the oldest records of beekeeping in the form of paintings by prehistoric man in the rock shelters. With the development of civilization, honey acquired an unique status in the lives of the ancient Indians. They regarded honey as a magical substance that controlled the fertility of women, cattle, as also their lands and crops. The recent past has witnessed a revival of the industry in the rich forest regions along the sub-Himalayan mountain ranges and the Western Ghats, where it has been practiced in its simplest form.

In India beekeeping has been mainly forest based. Several natural plant species provide nectar and pollen to honey bees. Thus, the raw material for production of honey is available free from nature. Bee hives neither demand additional land space nor do they compete with agriculture or animal husbandry for any input. The beekeeper needs only to spare a few hours in a week to look after his bee colonies. Beekeeping is therefore ideally suited to him as a part-time occupation. Beekeeping constitutes a resource of sustainable income generation to the rural and tribal farmers. It provides them valuable nutrition in the form of honey, protein rich pollen and brood. Bee products also constitute important ingredients of folk and traditional medicine.

The establishment of Khadi and Village Industries Commission to revitalize the traditional village industries, hastened the development of beekeeping. During the 1980s, an estimated one million bee hives had been functioning under various schemes of the Khadi and Village Industries Commission. Production of apiary honey in the country reached 10,000 tons, valued at about Rs. 300 million.

Side by side with the development of apiculture using the indigenous bee, Apis cerana, apiculture using the European bee, Apis mellifera, gained popularity in Jammu & Kashmir, Punjab, Himachal Pradesh, Haryana, Uttar Pradesh, Bihar and West Bengal. Wild honey bee colonies of the giant honey bee and the oriental hive bee have also been exploited for collection of honey.

Tribal populations and forest dwellers in several parts of India have honey collection from wild honey bee nests as their traditional profession. The methods of collection of honey and beeswax from these nests have changed only slightly over the millennia. The major regions for production of this honey are the forests and farms along the sub-Himalayan tracts and adjacent foothills, tropical forest and cultivated vegetation in Rajasthan, Uttar Pradesh, Madhya Pradesh, Maharashtra and Eastern Ghats in Orissa and Andhra Pradesh.

Resources and Potential

The raw materials for the beekeeping industry are mainly pollen and nectar that come from flowering plants. Both the natural and cultivated vegetation in India constitute an immense potential for development of beekeeping. About 500 flowering plant species, both wild and cultivated, are useful as major or minor sources of nectar and pollen. There are at least four species of true honey bees and three species of the stingless bees. Several sub-species and races of these are known to exist. In recent years the exotic honey bee has been introduced. Together these represent a wide variety of bee fauna that can be utilized for the development of honey industry in the country. There are several types of indigenous and traditional hives including logs, clay pots, wall niches, baskets and boxes of different sizes and shapes. In modern beekeeping, the combs are built on wooden frames that are moveable. This facilitates inspection and management of bee colonies. Three types of moveable frame hive are in common use : the Newton type along with its standardized version ISI Type A, the Jeolikote Villager, and its counterpart ISI Type B, and the Langstroth type. Besides the hives, the beekeepers need equipment and implements like the hive stand, nucleus box and smoker. The industry also needs equipment and machinery for handling and processing of honey, beeswax, for manufacture of comb foundation sheets, and for other operations.

India has a potential to keep about 120 million bee colonies, that can provide self-employment to over 6 million rural and tribal families. In terms of production, these bee colonies can produce over 1.2 million tons of honey and about 15,000 tons of beeswax. Organized collection of forest honey and beeswax using improved methods can result in an additional production of at least 120,000 tons of honey and 10,000 tons of beeswax. This can generate income to about 5 million tribal families.

Apiculture Technology

Production of honey has been the major aim of the industry. Modern beekeeping also includes production of beeswax, bee collected pollen, bee venom, royal jelly, propolis, as also of package bees, queen bees and nucleus colonies. All these are possible only with a proper management of bees, utilizing the local plant resources and adapting to the local climatic conditions. Modern beekeeping makes heavy use of beekeeping equipment and honey processing plant. This results in high efficiency and also ensures the quality of the processed honey.

Seasonal management of bee colonies varies in different parts of the country although the basic management methods are the same. Flow management, dearth management, provision of feeding, and control and cure of bee disorders, bee diseases, pests and enemies, are some of the routine measures to keep bee colonies healthy and strong. There are special management techniques like queen rearing, migration for honey production or for colony multiplication, which the beekeeper takes up after he gains sufficient knowledge and experience in handling bee colonies.

Market

About 10,000 tons of forest honey are produced annually. Apiary honey produced under the KVI sector is estimated to be a little less than 10,000 tons in 1990-91. Over 95 per cent of this was from the A. cerana colonies, the rest being from the European bee colonies. Forest honey, mostly from rock bee hives, is usually collected by tribals in forests and is procured by forest or tribal corporations as a minor forest produce. Quite a large quantity is also collected by groups or individuals on their own. Forest honey is usually thin, contains large quantity of pollen, bee juices and parts, wax and soil particles. The honey collector gets between Rs. 10 and Rs. 25 per kilogram of the forest honey. Forest honeys are mostly multifloral.

Apiary honey is produced in bee hives and is harvested by extraction in honey extractors. Other types of beekeeping equipment like queen excluder, smoker, hive tool, pollen trap, honey processing plant are also used. Beekeepers sell the honey to the co-operative society, if one exists in the area. In many parts of India, the beekeeper gets a much higher price if he sells it directly to the consumer. Apiary honeys are usually multifloral when marketed by state-level marketing organizations, because honeys from different sources are mixed while pooling, storage and processing. Several unifloral honeys are available in markets restricted to small areas within the state where it is produced. Rubber plant contributed to over 60 per cent of the total apiary honey production during 1990-91. Besides this, jamun, hirda, beheda, arjun, neem, litchi, palmyrah palm, eucalyptus, lagerstroemias, tamarind, cashew tree, scheffleras, tun, karanj, false acacia, wild shrubs like shain, crops of different varieties of mustards, sesame, niger, sunflower, berseem clover, khesari, coriander, orchard trees including different types of citrus, apple, puddum, cherry and other temperate fruit trees, coconut tree and coffee plantations are some important sources that provide unifloral honey.

Much of the forest honey is sold to the pharmaceutical, confectionery and food industries, where it is processed and used in different formulations. Apiary honey is usually processed at the producers level. This consists mainly of heating the honey and filtering. A few beekeepers or honey producers co-operative societies have better processing facilities that involve killing of honey fermenting yeasts. About 50 per cent of the apiary honey under the KVI sector is graded and marketed under AGMARK specifications.

In 1985 the consumption of honey was estimated to be about 8.4 g per capita, while in other countries this was 200 g. Presently this would be about 2.5 g. Honey has so far been consumed mainly as a medicine and for religious purposes. A small quantity has been used in kitchen as an ingredient of pickles, jams and preserves. With the increasing production in recent years, there is an increasing trend to use honey in food. This is obviously the case with the affluent segments of the population. Forest honey is used in pharmaceutical, food, confectionery, bakery and cosmetic industries.

One often finds a good demand for local honeys like honeys from Mahabaleshwar. People in Maharashtra have a strong liking for jamun, hirda or gela honeys which have acquired special individual medicinal significance. Similarly, kartiki honey in Kumaon, Uttar Pradesh is locally much favored. Some honeys have an essentially non-local market. Rubber tree honey can only be sold in non-local markets. Coorg honey with its characteristic flavour was well-known during 1950s and 1960s. Shain or sulah honey from Kashmir has been very popular.

Presently litchi honey from Bihar and other northern states is in great demand. The price structure is regulated by the market forces of supply and demand. Beekeepers in well-known hill stations and other places of tourist attraction take advantage of the popularity of honey and can market their produce at remunerative prices.

Indian honey has a good export market. With the use of modern collection, storage, beekeeping equipment, honey processing plants and bottling technologies the potential export market can be tapped.

BEE KEEPING INDUSTRY

Honey and beekeeping have a long history in India. Honey was the first sweet food tasted by the ancient Indian inhabiting rock shelters and forests. He hunted bee hives for this gift of god. India has some of the oldest records of beekeeping in the form of paintings by prehistoric man in the rock shelters. With the development of civilization, honey acquired an unique status in the lives of the ancient Indians. They regarded honey as a magical substance that controlled the fertility of women, cattle, as also their lands and crops. The recent past has witnessed a revival of the industry in the rich forest regions along the sub-Himalayan mountain ranges and the Western Ghats, where it has been practiced in its simplest form.

In India beekeeping has been mainly forest based. Several natural plant species provide nectar and pollen to honey bees. Thus, the raw material for production of honey is available free from nature. Bee hives neither demand additional land space nor do they compete with agriculture or animal husbandry for any input. The beekeeper needs only to spare a few hours in a week to look after his bee colonies. Beekeeping is therefore ideally suited to him as a part-time occupation. Beekeeping constitutes a resource of sustainable income generation to the rural and tribal farmers. It provides them valuable nutrition in the form of honey, protein rich pollen and brood. Bee products also constitute important ingredients of folk and traditional medicine.

The establishment of Khadi and Village Industries Commission to revitalize the traditional village industries, hastened the development of beekeeping. During the 1980s, an estimated one million bee hives had been functioning under various schemes of the Khadi and Village Industries Commission. Production of apiary honey in the country reached 10,000 tons, valued at about Rs. 300 million.

Side by side with the development of apiculture using the indigenous bee, Apis cerana, apiculture using the European bee, Apis mellifera, gained popularity in Jammu & Kashmir, Punjab, Himachal Pradesh, Haryana, Uttar Pradesh, Bihar and West Bengal. Wild honey bee colonies of the giant honey bee and the oriental hive bee have also been exploited for collection of honey. Tribal populations and forest dwellers in several parts of India have honey collection from wild honey bee nests as their traditional profession.

The methods of collection of honey and beeswax from these nests have changed only slightly over the millennia. The major regions for production of this honey are the forests and farms along the sub-Himalayan tracts and adjacent foothills, tropical forest and cultivated vegetation in Rajasthan, Uttar Pradesh, Madhya Pradesh, Maharashtra and Eastern Ghats in Orissa and Andhra Pradesh.

INDIGENOUS BEE KEEPING

The Challenge

Agricultural growth rates have fallen over the last two decades despite it being the main source of livelihoods in rural India. One of the major reasons in cross pollinated crops is reduction in the beneficial insect population. As per the studies, 95% of cross pollination happens through insects like honey bees, butterflies, moths etc., Due to indiscriminate usage of pesticides over a period decrease in the beneficial insect population is observed which leads to reduction in pollination percentage and yield.

The Solution

Honey bees are found the efficient pollinators compared to any other beneficial insect. Common honey bees used for bee keeping are indigenous bee and European bee. European bee is vulnerable to Indian climate and several diseases and requires high maintenance but the Indian indigenous bee is practically resistant to the popular disease like Thai sac brood. UTMT is encouraging farmers to go for indigenous bee keeping in the agricultural fields which increases the crop yield and requires low maintenance. Apart from an increased yield, farmers will also get additional income by selling the honey produced by the bees.

The Innovation

Under The Mango Tree (UTMT) seeks to position homestead bee keeping not just as a vehicle to diversify livelihoods for India's small and marginal farmers and increase their incomes by creating a sustained and direct market access, but for the larger and more important role of increasing agricultural productivity across India. By marketing the honey collected, UTMT also provides farmers with direct and sustained market access. Upon profitability, UTMT will distribute profits to improve the quality of life for farmers and their families.

To support implementation of the model, UTMT created a hybrid institutional structure that is symbiotically linked to invest in improving lives of poor producers:

- For profit company - to market and sell produce of poor primary producers directly to consumers. (UTMT is registered as a private limited company.)

- Not-for profit Society - to addresses livelihood diversification, agricultural productivity and training and capacity building amongst poor primary producers. (UTMT Society is registered under the Societies Registration Act, 1860.)

1. To increase productivity of crops in the State by providing honey bees to the farmers for pollination.
2. To maintain ecological balance in nature by way of domestication of honey bee species.
3. To maintain small apiaries for demonstration, pollination, extraction and popularisation of honey and other by-product of beekeeping.
4. To encourage farmers' participation in scientific beekeeping.
5. To provide technical know-how to the farmers of the State and organisation of short duration training courses in beekeeping.
6. Motivation of unemployed youth to adopt beekeeping as source of their livelihood.

The modern beekeeping in Himachal Pradesh was introduced only in the year 1934 in Kullu valley and in 1936 in Kangra valley. Only Apis cerana indica the Indian honeybee was reared in the State until the year 1961 when Apis mellifera from Italy was introduced in the State at Bee Research Station, Nagrota in Kangra. Keeping in view the importance of beekeeping in fruit industry the scheme was transferred from Agriculture to Horticulture department at time of bifurcation of the Agriculture department during 1971.

Prior to April, 1971 there were only 1250 bee colonies managed in modern bee hives in whole of Himachal Pradesh. After the scheme was transferred to the Horticulture department in the year 1971, the progress in the field of beekeeping has increased manifolds.

In view of the importance of honey bees in the successful pollination of fruit plants and the production of valuable honey and the bees wax, the department of Horticulture H.P. is taking up this scheme on priority basis and is making head way progress.

At present the department has 32 beekeeping demonstration apiaries at various suitable places in the state. The whole state is divided into two zones i.e. North and South zone. South zone has 17 demonstration apiaries whereas North zone has 15 demonstration apiaries. Prior to 1983-84 Apis cerana indica bee colonies were being maintained at all the beekeeping stations in South zone and Apis mellifera was only restricted to North zone.

After1983-84 when the Apis cerana indica bee colonies were almost wiped out by the epidemic caused by Thai sac brood virus and presently, over 90% of the bee colonies owned by the department and almost all the bee colonies managed by the private beekeepers now belong to the Apis mellifera species except the bee colonies kept in wall hives. The bee colonies are migrated to the other states by the beekeepers particularly during winters due to scarcity of bee flora to the colonies at that time in the State.

As a matter of fact Apis mellifera performance in Himachal Pradesh has been so encouraging that at present nearly 1500 unemployed educated youths have taken to Apis mellifera beekeeping as full time profession and are maintaining nearly 80000 bee colonies as against 4200 bee colonies in the year 1981-82. These bee colonies are capable of producing about 1600 M.T. of honey annually as against 3 M.T. during 1981-82.

Importance of honey bees in pollination of fruits plants needs no emphases. It is an established fact that the value of increased fruit harvests as a result of honey bee pollination is 14 to 20 times more than the value of honey obtained directly from the bee colonies. The department helps the farmers by arranging bee colonies for pollination to the orchardists at nominal rent. According to the modest estimates about 2,00,000 bee colonies are needed for appropriate pollination for the bearing orchards in the state.

HOLISTIC BEEKEEPING

New (old) hive designs

Once we approach bee keeping in the context of the "Bien", which represents the undividable entity of the hive, our methods and hives will change accordingly. There are currently three alternative hive designs (and more to come) used at "The Melissa Garden":

One-Room-Hive (Golden Hive)

One of the "bee-natural" hives we work with at the Melissa Garden is the "one-room-hive" (in German: "Einraumbeute"), which is also called the "golden hive". It is designed to provide the best environment for the development of the "Bien" and to minimize necessary manipulation .

Four different elements are part of the new design:

- The entire colony lives in one room (without multiple hives and frame levels)
- The hive comes with tall frames. That size comb sustains the "Bien" and allows the development of a large brood nest.
- The side window enables the bee keeper to receive information about the cycle/status of the "Bien" without having to open the hive. A small size "indicator comb" can be build along side the viewing window.
- The dimensions of the one-room-hive are set according to the "golden mean". It is a universal principle within all forming forces in nature and is found in art, architecture and ancient philosophy. It's also called the "divine proportion".

It was designed by Mellifera e.V., the German holistic bee keeper association. The "Golden Hive" provides an environment for bees that is closer to their natural gestalt. It gives the bees the space to build natural comb with greater depth than regular bee hives. The brood nest is a protected space, and honey can be received from the sides. This hive contains 20 frames and is not supered.

The comb surface area equals the frames of two regular deep and one medium Langstroth hive bodies. It has the typical screened bottom board for varroa monitoring, and uses follower boards to support changing bee populations throughout the season. A wax cloth lays on top of the frames and provides further options for protection the inner climate of the "Bien".

"Weissenseifener Haengekorb" (Round Skep Hive)

The "Weissenseifener Haengekorb" was designed by the German sculptor Guenther Mancke. The form and shape of the hive are created according to natural/wild bee hives. The "Bien" as "an organic interpretation of an individual" (Tautz) was the blueprint for the design. Already through his outer shape it reveals the nature of the bees colonie– as if the egg shaped skep would be the outer shell or skin of this living being.

The inner shape allows bees to unfold their own natural gestalt, in harmony with their instinctual life forces. The "Haengekorb" is made out of rye straw and has nine, half moon shaped arched, movable frames. Comb is built naturally and can be almost 2 feet deep. Supering is possible while fully protecting the integrity of the brood nest. The entrance is located at the bottom of the hive.

Top Bar Hive

It provides all the features of a natural comb bee hive. Top bars fully cover the upper opening of the hive, with initial comb guidance on the lower bar side. Top bar hives are used in many different cultures. We are introducing new versions of the top bar hive this coming year, which will use "bee-natural" hive proportions and will provide more space for larger comb creation. Top bar hives can be built easily with some basic materials.

Langstroth hives with natural comb

All of the Langstroth hives at "The Melissa Garden" allow bees to live on natural comb. Deep hive bodies are added from below (no supering). Bees over-winter on their own food and only true surplus honey is harvested.

Principles of Holistic Beekeeping

Interconnectedness

There is no single bee – as there is no single human being. It's a product of a limited world view. The single bee is only one individual part of the bigger entity of the entire bee hive. The "Bien" is what Tautz calls an "organic interpretation of an individual". This can serve as a beautiful metaphor or mirror for our own existence: That we are just one individual part of a bigger entity: The earth's ecosystem and the entire universe.

Shift Paradigms

Bees are an indicator species, reflecting the health and status of our life environment as well as the interdependency and interconnectedness of all life on earth. Traditional beekeeping and farming understood and acknowledged the natural life forces of the bees. Modern beekeeping and farming practices have lost this ancient knowledge and this loss has taken its toll on the bees on multiple levels. We have to shift paradigms in order to save the bees.

"Bien"

The concept of the "Bien" describes the undividable entity of the hive. The whole is one organism and the hive is more than the sum of the individual parts. Thousands of bees are integrated into a higher-order entity, one whose abilities far transcend those of the individual bee. "The consciousness of the beehive (not of the individual bees) is of a very high nature" (Rudolf Steiner).

Their communication and networking capacities, non hierarchical decision processes and an understanding of service to the greater web of life, which the individual being (bee) is part of, are pointing to a higher level of development and awareness. And such, the bees are a vital part of human culture and an inspiration to the soul. Being in touch with the "Bien" also means to reach out to the flowering world. As bee-keepers we are becoming "flower-keepers" and stewards of the earth as well.

Comb

The comb co-evolved with the bees as a part of the bee itself. Wax comb is the biggest inner organ of the "Bien". Bees spend 90% of their life on the comb. They create the wax out of their own body – no other insect is able to do this. The comb is home, womb, pantry, (external) skeleton, sense organ, nerve system, memory organ and immune system. The "comb-wide-web" provides a means of communication on multiple levels: dance, vibration, chemical marking.

The dance floor is marked with some bee pheromones and other still unknown substances. The comb is a controlled environment. As the interior milieu, it becomes part of the "Bien". Therefore, it is essential to allow bees to build their own natural comb and to give them the freedom to express their instinctual life forces. Natural comb is essential when we want to support the bees in a time of ecological challenges. It is their birthright.

It is easy to let bees build natural comb. Since bees build according to gravity, hives need to be leveled and frames need to provide some initial guidance, such as beads of wax across the bottom of the top bar, or one-inch strips of bees wax foundation placed for the bees to start building from. Tapered frames may be used as well, since bees draw naturally comb from thin edges. If you have questions in regards to natural comb management, please contact us for assistance.

NEST DENSITY – THE LANDSCAPE BECOMES THE APIARIUM

Following the movement of the "Bien", swarms aim to settle further away from their mother colony. The natural distribution of nest sites vary according to climate and local flora. It is a natural instinct of the "Bien" to leave home and journey into the landscape. It not only reflects a consideration for forage, but it also serves the health of the bees by favoring vertical over horizontal transmission of so called pathogens.

The former leads to a lower virulence of diseases where the latter leads to an unbalance of relationships, in particular within the symbiotic life between the "Bien" and it's microorganisms and parasites like varroa. Therefore the landscape shall become the apiarium again. Hives are set up individually and spaced at least 400 yards from each other. Various hives are set up on tables and in trees within the boundaries of the Melissa Garden. Empty hives will be "seeded" by swarms only, whether feral or from Melissa Garden hives. We hope to help restore the natural web of life this way on various levels.

Parallels between "Bien" and mammals

There are interesting parallels between bees and mammals. Both have low reproduction rates in common. Mammals raise their offspring with mother's milk and nursing bees use "sister's milk," which is produced in special glands. Wherein mammals provide a uterus, the brood nest of the bees has similar characteristics as a "social" uterus. Body temperatures are 98 degrees F for humans and 95 degrees F for bees, which is very close.

Retention of nest scent and heat

The hive is not an external dwelling of the "Bien". When we open a bee hive, we are entering an ecological system, or even a being's body. Heat, humidity, light, draft, the entire self awareness, the immune system and the sense of integrity of the "Bien" are challenged and affected! Johann Thür called the internal conditions "the element of life, the retention of nest scent and heat", which are part of the immune system and important for it's well being (studies have shown that more frequent opening of hives caused higher damage due the hive beetle).

Approach

When opening a h]ive, how do we approach the "Bien"? It seems we are conditioned to reach for beekeeping tools like a veil, gloves and smoke. Our "armor" may make it challenging to stay open for the mood of the bees and it may be more difficult to extend our empathy. With the protection in place, we more likely penetrate realms of highest vulnerability, without even noticing. We may get out of touch with what is right in front of us. To re-evaluate our precautions and body protections (which are important), we may open up to a new approach of respectful encountering the "Bien". We can learn to listen and how to move and "dance" with the bees. This does not mean not to protect ourselves. The issue is rather not to become "blind" through our "beekeeping tools". Smoke is another stress factor in beekeeping, and with all our senses open, the smoker can stay cold for most of the time. Experiment with it and go with you comfort level. Then a different path of beekeeping opens up.

Mumbai tunes into urban bee-keeping buzz

Soon, Colaba and Capitol Hill could have something in common. The city is the latest to join the elite company of London, Paris, San Francisco, Washington, Toronto and Chicago, among others, by embracing the trend of urban bee-keeping. For the first time in India, an urban bee-keeping training programme is being conducted at different locations in the city by Under The Mango Tree, an NGO. "The need for such a programme arose when we realised that there is little awareness, especially in urban India, about bees and their crucial role in the environment," said Sujana Krishnamoorthy, programme head, UTMT. "Bees are nature's most prolific pollinators and a way for plants to continue their lineage. A majority of the fresh fruits and vegetables we eat rely on pollination by bees," said Gurushabd Khalsa, co-ordinator for the programme. Technical experts from Uttarakhand have been roped in to conduct the training. "We will use apis cerana indica bees, which are locally available and most suitable for Mumbai's environment. The bees' home will be a box made of wood and has frames where the bees build their honey comb.

2

Origin and Development of Apiculture

Bees and honey have featured extensively in culture and mythology throughout history. Cave drawings found in Spain dated around 6000 BC depict honey gatherers climbing trees to harvest honey from wild bees. These are thought to be some of the earliest illustrations of humans risking life and limb to obtain honey. Similar drawings have also been found in India, Africa, Asia and Australia.

While beekeeping dates back at least 4, 000 years there are still many parts of the world where honey is gathered by climbing trees or rock faces. One thing we can be sure of is that humans have always had a sweet tooth and would go to great lengths to obtain honey.

Bees and honHendingey have always been associated with fertility and had a special place in ancient Greece where the great mother Goddess Artemis was depicted as half woman and half bee.

The Mayans too were expert beekeepers and the Spaniards who came across them in the 16th century were surprised to find apiaries containing one to two thousand hives.

HONEY AND WAX

Honey was held in such high regard that it was presented as an offering to the Gods as ambrosia. The medicinal properties of honey were well known and it was used as an antiseptic and preservative. Honey was also included as an ingredient in creams, lotions and alcoholic drinks. Wax and honey were used in the mummification process and jars containing honey have been found in Egyptian pyramids. A story is told of a jar being opened and the finders tasting the contents and finding it still edible, until some hair was found in the honey!! On further investigation the jar was found to contain the fully preserved body of a child!!

Another important product harvested from the hive is wax. Candles made from beeswax were long lasting and had little smell and became essential in church services. Monks were able to meet the constant demand for candles by becoming beekeepers.

One of the most renowned Beekeeper monks was Brother Adam of Buckfast Abbey who developed a strain of bee valued the world over for its good temper and prolific honey production.

Skeps and Boles

In medieval times bees were usually housed in hollowed-out logs or in skeps, and placed in alcoves in walls called bee boles. The word 'skep' is thought to originate from the old Norse word 'skeppe' which was used to mean a basket.

Harvesting the honey involved killing the colony by smoking with burning sulphur, and shaking the dead bees out of the skep. Boles can still be seen in old walls, though now may have a variety of uses. Some are used for garden ornaments, or potted plants, and the larger ones (bee shelves) have been used as wood stores. Generally the recesses in stone walls were 18-30 inches (46-76 cm) in height and 15-26 inches (38-66 cm) wide and 14-21 inches (36-53 cm) deep.

IBRA (International Bee Research Association) has records of bee boles and other beekeeping structures built in the past. Started by Dr Eva Crane in 1952, the Register now contains paper records for 1540 sites, including many in Wales.

Modern Beekeeping

There are many 'greats' in the world of beekeeping (each country can claim to have its own master beekeeper) who advanced the craft in some unique way. For example Rev. Charles Butler (1559-1647) described the functions of bees within the hive, how to manage swarms and honey collection in 'The Feminine Monarchy' (1623). Before this it was widely believed that the worker bees were all male. Thomas Wildman (1734-1781) developed the method of layering skeps, one on top of another. In this way the honey could be harvested without killing the bees. And

Francois Huber (1750-1831) invented the 'leaf' hive making it possible to inspect frames, and identified the life cycle of bees.

However modern beekeeping, where bees are kept on moveable frames contained in wooden boxes, started about 150 years ago. Although many types of hive were designed in the 18th and 19th centuries, it is the American born Reverend L. Langstroth who is credited with the invention of an effective moveable frame hive, and over 70% of hives in use today are based on his design.

Bees in Wales

Conwy has a long association with beekeeping, dating back perhaps 1, 000 years, and bee boles can be found in walls throughout the Conwy Valley, though there must be many more yet to be discovered.

In Conwy the holding of the annual Conwy Honey Fair is granted within Conwy Town's Royal Charter and was granted by Edward I over 700 years ago. It is recorded that Edward bought a cask of honey at the fair in the 13th century.

The Honey Fair takes place every September on Conwy High Street, where beekeepers sell honey, honey soaps, beeswax candles and polish. Local beekeepers can sell about a tonne of honey by lunchtime!

In 2011 the National Trust undertook the restoration of a wall of bee boles at Dolmelynllyn Hall at Ganllwyd near Dolgellau. Prior to restoration the wall was officially recorded, 10 years ago, as having 38 boles, but as the work progressed a further 8 boles were found, making it the largest wall of bee boles in the UK. In May 2012 the Merioneth Historical and Record Society and Merioneth Beekeepers Association began a joint project to identify bee boles in the Merioneth area. So far from 9 previously identified sites, they have now discovered a further 2 exceptional sites with bee shelves, and have more hitherto unknown sites to investigate.

The Bee Hive

Prior to the middle of the 1800's, most bee hives in North America and Europe were simple shelters for the bees. Skeps, log gums and box hives were common types of hives in this period.

Bees attached their wax combs to the hive's roof and walls, just like they do in wild hives. Today we refer to these types of hives as fixed-comb hives.

Skeps were made from grass straw, and often had sticks inside to provide support for the honey combs. Beekeepers inspected skep hives from the bottom.

Box hives were simple shelters to house a swarm of bees.

Log gums were made from hollow logs, fitted with a roof.

Sometimes a box or container was added on top of a log gum or box hive for the bees to store honey.

It was also hard to get honey from these hives without damaging or destroying the bee colony and getting the bees upset (they sting, you know!).

Some hives in the 1800's used clever designs that discouraged queens from laying eggs in some parts of the hive, so honey could be harvested without damaging the colony. These beekeepers knew that queens tended not to lay eggs in more than one area in the hive, so they made side and top compartments with passageways for the bees.

The hives shown here have a place for the brood nest in the center, and places for honey storage on the sides. This is a kind of queen excluder that relied on the behavior of bees instead of a physical barrier. Today, we know that pheromones influence organization within a bee hive

Some skeps and box hives from the 1800's also had a second container, or "super" for the bees to store honey such as the one at right.

The "Nutt Collateral Hive" at right is a particularly fancy hive that used the concept of a pheromone-based queen excluder. The use of supers and separate honey compartments allowed the beekeeper to remove honey without destroying the colony.

Supers were sometimes put on top of log sections, or "gums", so that honey could be harvested easily.

In these hives, it was hard to know when the bees had a problem with disease, or when they became queenless or were starving. The beekeeper could not inspect each comb to see what was wrong.

Fixed-comb hives like the ones above were popular until the 1850's, and yielded 10-15 pounds per colony each year, according to Root's ABC book from 1895. Of course, many things helped increase honey yields since then, including the Italian bee.

It was known for a long time that bees liked to build their honey combs about 1 and 3/8 inches apart. Honey comb is about one inch wide, so this left a 3/8 inch passageway between the combs.

Some beekeepers built hives that forced the bees to build combs along "top bars" that were spaced about 1 and 3/8 inches apart. Top bars allowed the beekeeper to carefully remove combs for inspection without damaging them. These are called movable comb hives. This hive from Greece in the 1600's (right) uses this concept.

Movable comb hives allow beekeepers to start new colonies easily by dividing a hive. They also allow beekeepers to inspect the health of colonies, find the queen, and even cut honey comb without destroying the brood nest. Bees in movable comb colonies were disturbed less than bees in fixed-comb hives, so beekeepers received fewer stings!

Many movable comb hive inventions used "frames" for the bees to build their combs inside.

The Leaf Hive, invented in Switzerland in 1789 by Francis Huber, was a fully movable frame hive. The combs in this hive were examined like pages in a book. A. I. Root and E. R. Root credit Huber with inventing the first movable *frame* hive. Huber's contribution was also acknowledged by Lorenzo Langstroth, inventor of the hive style that is most commonly used today:

"The use of the Huber hive had satisfied me, that with proper precautions the combs might be removed without enraging the bees, and that these insects were capable of being tamed to a surprising degree. Without knowledge of these facts, I should have regarded a hive permitting the removal of the combs, as quite too dangerous for practical use. "

- L. L. Langstroth in *Langstroth on the Honey-Bee*, 1860.

The Quinby closed-end frame hive had many good features of a movable-frame hive. The side bars of the Quinby frame (at right) also formed the walls of the hive. Some successful beekeepers were using this hive as late as the 1890's.

Skeps, log gums and box hives remained most common style of bee hives, despite these movable comb inventions in the 1700's and early 1800's.

A major improvement in hive design was made in 1851 by Lorenzo Langstroth. He built a hive with frames that hung from the top ends of the hive, leaving a 3/8 inch space between all sides of the frames and the hive body.

His clever design used the principle that bees usually do not build comb in 3/8 inch passageways. If the space is bigger than 3/8 of an inch, the bees will build comb. If it is less than 1/4 inch, they will attach propolis.

Langstroth's frames were easily handled without breaking the comb. Today we refer to the 3/8 inch passageways as a "bee-space. " This practical hive is the direct ancestor of the modern hive that is most popular today.

In describing the benefits of his hive with movable frames, Langstroth wrote:

"... the chief peculiarity in my hive was the facility with which they could be removed without enraging the bees.... I could dispense with natural swarming, and yet multiply colonies with greater rapidity and certainly than by the common methods.... feeble colonies could be strengthened, and those which had lost their queen furnished with the means of obtaining another..... If I suspected that any thing was wrong with a hive, I could quickly ascertain its true condition, and apply the proper remedies. "

- L. L. Langstroth in *Langstroth on the Honey-Bee*, 1860.

By the 1890's, the movable-frame hive was largely adopted for general use. Most of the hives in the late 1800's used Langstroth's bee-space frame concept. Many of the hive designs were more elaborate than ones used today.

Both 8-frame and 10-frame hives were popular. Some hives had double-thick walls filled with wheat chaff for insulation.

By the year 1900, most modern beekeepers were using variants of the Langstroth hive with Hoffman-style frame, like the ones used today.

These inventions helped make beekeeping a viable business.

The modern bee hive has not changed very much during the 20th century. The most significant beekeeping advances of the 20th century involved the extracting process and bee management.

The evolution of the bee hive will surely continue. Can you guess what the next 100 years will bring?

BEE HUNTING

Fig. 1

Glass-topped boxes were used to find wild bee colonies. The wild colonies would be raided for the honey, and the bees sometimes put into a bee hive.

Bees were tempted into the box with a sweet syrup or honey, and the hunter followed the bees as they flew back to their colony. A. I. Root described how to use the hunting box:

"The only season in which we can trap bees is when they will rob briskly at home... Cautiously set the box over the first bee you find upon the flowers. As soon as the box is well over the flower, close the bottom with your hand, and the bee will buzz up against the glass. Catch as many as you wish, in the same way, and they will soon be sipping the honey. Before any have filled themselves, ready to fly, place your box on some elevated point, such as the top of a stump in an open space in the field, and draw back the glass slide....

"When a bee comes back, you will recognize it by the peculiar inquiring hum, like robbers in front of a hive where they have once had a taste of the spoils.... As soon as you are pretty well satisfied in which direction they are located, you can close the glass slide and move along the line, nearer the woods. Open the box and you will have them just as busy again; mark the line and move again, and you will soon follow them to their home. To aid you in deciding where they are, you can move off to one side and start a cross-line. Of course, the tree will be found just where these lines meet. "

Swarms

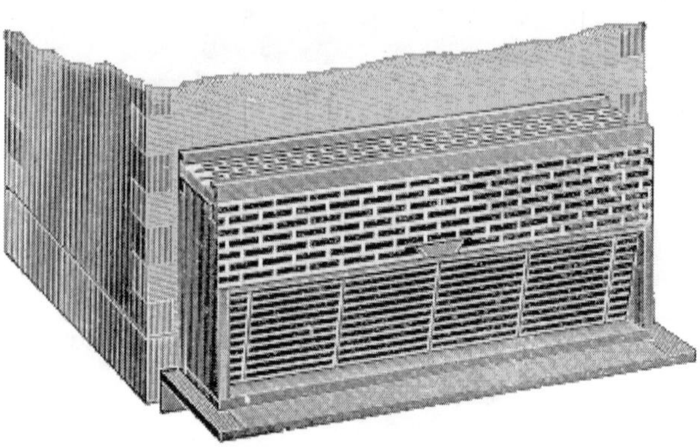

Fig. 2

Until the late 1800's, catching swarms was a primary way of increasing one's colonies. Several devices were designed to help stop or catch swarms as they left the hive.

The basic concept in most of these devices was to direct a swarm into another hive or enclosed area. There also devices designed to stop queens from leaving the hive.

Queen Rearing

Grafting of worker larva to raise queens began in the late 1800's. Before that, some books described how to cut worker comb containing eggs and place it in a queenless hive.

Fig. 3

Shipping bees through the postal system has not changed much during the past 100 years. In the 1890's, Italian and Carniolan queens cost about $1 each. The black bee that had been popular in earlier years cost about 25 cents. Hybrids cost about 40 cents. Albino queens were also available.

Bee Diseases & Parasites

Foulbrood was the most feared bee disease.

The wax moth was accused of destroying healthy bee colonies. Today we know that the wax moth is almost always controllable by a healthy bee colony, and that its presence in a hive is usually a symptom of another problem. One creative beekeeper's way of controlling wax moth follows below, quoted from a book titled "Moore's Universal Assistant" (1878):

"HENS MADE TO PROTECT BEES. - A bee raiser has patented an invention for the protection of bees from the attacks of the honey bee moth, which enters the hives at night, and rifles the stores.... Hens, he observed, retire to rest early; but bees seek repose earlier still; no sooner are they sunk into slumber, than the moths steals into their abode and devours the produce of their toil. He has now built a stand of hives with a hen house connected. The bees first betake themselves to their dwelling and settle themselves for the night. The hens then come home to roost on their perch, and as they take their places upon it, their weight sets some simple mechanism to work, which at once shuts down the doors of all the hives. When the day dawns, however, the hens leave their roost, and the removal of their weight from the perch raises the hive doors, and gives egress to the bees in time for their morning's work. "

Oil pots with flaming wicks were also used to attract and kill moths.

Wintering

Wintering hives in field shelters or cellars seemed popular in the northern U. S. One magazine article suggested 40-45 degrees f to be the ideal temperature for wintering bees in a cellar.

Smokers came in a variety of designs.

"We can drive cattle and horses, and, to some extent, drive even pigs, with a whip; but one who undertakes to drive bees in any such way will find to his sorrow, that all the rest of the animal kingdom are mild in comparison"

-A. I. Root, in The ABC of Bee Culture, 1895

Some beekeepers used a technique called "drumming" to remove bees from a hive. They would put an empty box tightly on top of the hive and close up the hive entrance. Then they would beat the sides of the hive like a drum for 10 or 15 minutes. All the bees would move up to the empty box, leaving the hive empty so the honey could be removed.

Veils

The veils were practical, though somewhat different from veils today.

Moving Bees & Honey

Bees were moved on horse-drawn trailers. Some horses even had bee suits to protect them from stings!

Hive staples apparently were not invented yet. Books recommended tying hives together with rope when transporting them.

The Honey House

Uncapping

Steam and electric uncapping knives appeared around the turn of 1900. The steam or electricity kept the knife warm, allowing it to easily cut through beeswax. Before heated knives, beekeepers often used several knives, using one while another was being heated.

The upcapping table at right (1913) used a burner below to melt the cappings and separate the wax and honey. It also had a warming pan to heat the knives (far right).

Extracting

Small-scale honey extractors haven't changed much during the past 100 years. The centrifugal extractor was invented in 1865 by an Austrian, Major Hruschka.

Fig. 4

The one-frame pull-cord design (at right) was a clever idea. Bohn's extractor used an early radial design.

Fig. 5

The radial extractor, where both sides of frames are extracted at the same time, was a time-saving invention. The early radial extractor at right appears in a book dated 1895.

This wagon extracting setup shows a process very similar to that used today. It may have been used at night, when the bees' robbing wouldn't be a problem.

This wagon extracting setup shows a process very similar to that used today. It may have been used at night, when the bees' robbing wouldn't be a problem.

Fig. 6

Fig. 7

This drawing is from 1913, showing honey being separated from wax by using heat. This concept of drawing off honey from the bottom of a container and wax from the top is still used today.

Bottling

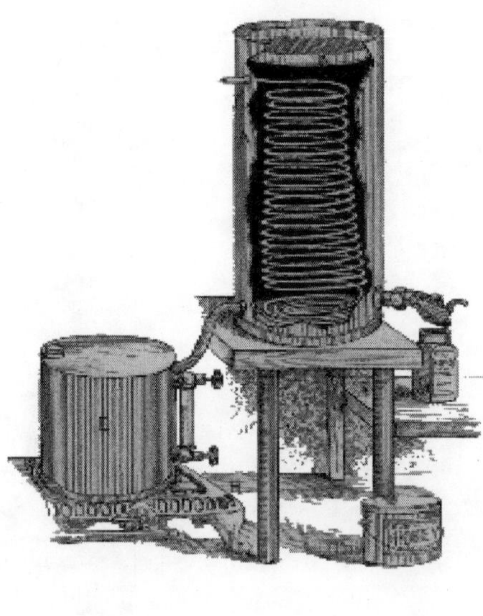

Fig. 8

In 1913, granulated honey in glass containers was considered unsalable, according to Root. To delay granulation, or crystalli-zation, of honey, heating it prior to bottling became popular. At right, steam was forced through coils inside a honey tank, warming the honey and delaying granulation.

Honey

Comb honey was also more popular than it is today. Buying honey in the comb was reportedly a way to avoid buying honey that was not pure.

Vinegar was also being made from honey, a use for honey that is rarely seen today.

Observation Bee Hives

Honey bees have fascinated people for centuries, and people have built many types of hives to observe them.

The woodcut at right shows a woman observing bees through a window in an elaborate observation hive.

The scene at left shows a Ukrainian observation hive from about 1840.

Thomas Nutt built an elaborate observation hive in the 1830's. The center section of his hive housed the brood nest, and the bees stored honey on either side. The hive also had a hollow wooden floor with drawers for feeding the bees.

Fig. 9

Fig. 10

Several of Langstroth's hives from the 1850's and 1860's had glass walls installed for observation. The Langstroth hive at right is fitted with many glass jars and bells for the bees to fill with honey. The bees would need to be very crowded before filling those jars!

Fig. 11

Fig. 12

Forcing bees to build comb inside glass jars is a common theme for observation hives.

The glass bell on the left has a wood frame inside for the bees to attach their comb. Without a wood frame, the bees will readily attach comb to the glass.

Fig. 13

Bees will build comb in glass jars on top of standard hives, too, if the bees are crowded. The hive at right is a fancy skep with three jars on top. The jars are normally covered.

The bees' natural comb-building, without frames, is very interesting. I highly recommend that curious beekeepers build an un-framed hive at least once for the learning experience. This hive on the left is a natural comb hive from the late 1800's.

Fig. 14

Alfred Neighbour built many observation hives in the 1870's, including the hive at left. The hive at right is a simple adaptation using standard frames and comb sections.

Neighbour also established an observation beehive building at a zoological garden for the public to view bees.

The wall-mounted hive at right has a well-designed entrance. It pivots against the wall when not being viewed.

APICULTURE IN INDIA

Honey and beekeeping have a long history in India. Honey was the first sweet food tasted by the ancient Indian inhabiting rock shelters and forests. He hunted bee hives for this gift of god. India has some of the oldest records of beekeeping in the form of paintings by prehistoric man in the rock shelters. With the development of civilization, honey acquired an unique status in the lives of the ancient Indians. They regarded honey as a magical substance that controlled the fertility of women, cattle, as also their lands and crops. The recent past has witnessed a revival of the industry in the rich forest regions along the sub-Himalayan mountain ranges and the Western Ghats, where it has been practiced in its simplest form.

In India beekeeping has been mainly forest based. Several natural plant species provide nectar and pollen to honey bees. Thus, the raw material for production of honey is available free from nature. Bee hives neither demand additional land space nor do they compete with agriculture or animal husbandry for any input. The beekeeper needs only to spare a few hours in a week to look after his bee colonies. Beekeeping is therefore ideally suited to him as a part-time occupation. Beekeeping constitutes a resource of sustainable income generation to the rural and tribal farmers. It provides them valuable nutrition in the form of honey, protein rich pollen and brood. Bee products also constitute important ingredients of folk and traditional medicine.

The establishment of Khadi and Village Industries Commission to revitalize the traditional village industries, hastened the development of beekeeping. During the 1980s, an estimated one million bee hives had been functioning under various schemes of the Khadi and Village Industries Commission. Production of apiary honey in the country reached 10, 000 tons, valued at about Rs. 300 million.

Side by side with the development of apiculture using the indigenous bee, Apis cerana, apiculture using the European bee, Apis mellifera, gained popularity in Jammu & Kashmir, Punjab, Himachal Pradesh, Haryana, Uttar Pradesh, Bihar and West Bengal. Wild honey bee colonies of the giant honey bee and the oriental hive bee have also been exploited for collection of honey.

Tribal populations and forest dwellers in several parts of India have honey collection from wild honey bee nests as their traditional profession. The methods of collection of honey and beeswax from these nests have changed only slightly over the millennia. The major regions for production of this honey are the forests and farms along the sub-Himalayan tracts and adjacent foothills, tropical forest and cultivated vegetation in Rajasthan, Uttar Pradesh, Madhya Pradesh, Maharashtra and Eastern Ghats in Orissa and Andhra Pradesh.

RESOURCES AND POTENTIAL

The raw materials for the beekeeping industry are mainly pollen and nectar that come from flowering plants. Both the natural and cultivated vegetation in India constitute an immense potential for development of beekeeping. About 500 flowering plant species, both wild and cultivated, are useful as major or minor sources of nectar and pollen.

There are at least four species of true honey bees and three species of the stingless bees. Several sub-species and races of these are known to exist. In recent years the exotic honey bee has been introduced. Together these represent a wide variety of bee fauna that can be utilized for the development of honey industry in the country.

There are several types of indigenous and traditional hives including logs, clay pots, wall niches, baskets and boxes of different sizes and shapes. In modern beekeeping, the combs are built on wooden frames that are moveable. This facilitates inspection and management of bee colonies. Three types of moveable frame hive are in common use : the Newton type along with its standardized version ISI Type A, the Jeolikote Villager, and its counterpart ISI Type B, and the Langstroth type.

Besides the hives, the beekeepers need equipment and implements like the hive stand, nucleus box and smoker. The industry also needs equipment and machinery for handling and processing of honey, beeswax, for manufacture of comb foundation sheets, and for other operations.

India has a potential to keep about 120 million bee colonies, that can provide self-employment to over 6 million rural and tribal families. In terms of production, these bee colonies can produce over 1. 2 million tons of honey and about 15, 000 tons of beeswax. Organized collection of forest honey and beeswax using improved methods can result in an additional production of at least 120, 000 tons of honey and 10, 000 tons of beeswax. This can generate income to about 5 million tribal families.

Apiculture Technology

Production of honey has been the major aim of the industry. Modern beekeeping also includes production of beeswax, bee collected pollen, bee venom, royal jelly, propolis, as also of package bees, queen bees and nucleus colonies. All these are possible only with a proper management of bees, utilizing the local plant resources and adapting to the local climatic conditions. Modern beekeeping makes heavy use of beekeeping equipment and honey processing plant. This results in high efficiency and also ensures the quality of the processed honey.

Seasonal management of bee colonies varies in different parts of the country although the basic management methods are the same. Flow management, dearth management, provision of feeding, and control and cure of bee disorders, bee diseases, pests and enemies, are some of the routine measures to keep bee colonies healthy and strong. There are special management techniques like queen rearing, migration for honey production or for colony multiplication, which the beekeeper takes up after he gains sufficient knowledge and experience in handling bee colonies.

Market

About 10, 000 tons of forest honey are produced annually. Apiary honey produced under the KVI sector is estimated to be a little less than 10, 000 tons in 1990-91. Over 95 per cent of this was from the A. cerana colonies, the rest being from the European bee colonies. Forest honey, mostly from rock bee hives, is usually collected by tribals in forests and is procured by forest or tribal corporations as a minor forest produce.

Quite a large quantity is also collected by groups or individuals on their own. Forest honey is usually thin, contains large quantity of pollen, bee juices and parts, wax and soil particles. The honey collector gets between Rs. 10 and Rs. 25 per kilogram of the forest honey. Forest honeys are mostly multifloral.

Apiary honey is produced in bee hives and is harvested by extraction in honey extractors. Other types of beekeeping equipment like queen excluder, smoker, hive tool, pollen trap, honey processing plant are also used. Beekeepers sell the honey to the co-operative society, if one exists in the area. In many parts of India, the beekeeper gets a much higher price if he sells it directly to the consumer.

Apiary honeys are usually multifloral when marketed by state-level marketing organizations, because honeys from different sources are mixed while pooling, storage and processing. Several unifloral honeys are available in markets restricted to small areas within the state where it is produced. Rubber plant contributed to over 60 per cent of the total apiary honey production during 1990-91.

Besides this, jamun, hirda, beheda, arjun, neem, litchi, palmyrah palm, eucalyptus, lagerstroemias, tamarind, cashew tree, scheffleras, tun, karanj, false acacia, wild shrubs like shain, crops of different varieties of mustards, sesame, niger, sunflower, berseem clover, khesari, coriander, orchard trees including different types of citrus, apple, puddum, cherry and other temperate fruit trees, coconut tree and coffee plantations are some important sources that provide unifloral honey.

Much of the forest honey is sold to the pharmaceutical, confectionery and food industries, where it is processed and used in different formulations. Apiary honey is usually processed at the producers level. This consists mainly of heating the honey and filtering. A few beekeepers or honey producers co-operative societies have better processing facilities that involve killing of honey fermenting yeasts.

About 50 per cent of the apiary honey under the KVI sector is graded and marketed under AGMARK specifications. In 1985 the consumption of honey was estimated to be about 8. 4 g per capita, while in other countries this was 200 g. Presently this would be about 2. 5 g. Honey has so far been consumed mainly as a medicine and for religious purposes. A small quantity has been used in kitchen as an ingredient of pickles, jams and preserves.

With the increasing production in recent years, there is an increasing trend to use honey in food. This is obviously the case with the affluent segments of the population. Forest honey is used in pharmaceutical, food, confectionery, bakery and cosmetic industries.

One often finds a good demand for local honeys like honeys from Mahabaleshwar. People in Maharashtra have a strong liking for jamun, hirda or gela honeys which have acquired special individual medicinal significance. Similarly, kartiki honey in Kumaon, Uttar Pradesh is locally much favored. Some honeys have an essentially non-local market.

Rubber tree honey can only be sold in non-local markets. Coorg honey with its characteristic flavour was well-known during 1950s and 1960s. Shain or sulah honey from Kashmir has been very popular. Presently litchi honey from Bihar and other northern states is in great demand. The price structure is regulated by the market forces of supply and demand. Beekeepers in well-known hill stations and other places of tourist attraction take advantage of the popularity of honey and can market their produce at remunerative prices.

Indian honey has a good export market. With the use of modern collection, storage, beekeeping equipment, honey processing plants and bottling technologies the potential export market can be tapped.

3

Apiculture Process

A smoker provides beekeepers with a source of cool smoke needed to calm the bees. The smoker consists of a fuel box containing smouldering fuel such as dried cow dung, hessian, maize cobs or cardboard, and has a bellows attached. The beekeeper puffs a little smoke near the entrance of the hive before it is opened and gently smokes the bees to move them from one part of the hive to another. Imported smokers are useful as prototypes, but smokers can be manufactured by village blacksmiths, which adds to local livelihoods.

PROTECTIVE CLOTHING

Abroad-brimmed hat with a veil protects the head and neck from stings. Protective clothing gives beginner beekeepers confidence; more experienced beekeepers find that wearing too much protective clothing makes them hot and the bulk makes it difficult to work sufficiently gently with the bees. Some beekeepers merely put a plastic bag over each hand, secured at the wrist with a rubber band. Rubber bands also prevent bees from crawling up trouser legs or shirtsleeves. Imported clothing can provide useful prototypes, but beekeeping clothes - basically modified overalls - can be made locally, thus providing a useful stimulus for local industry.

Hive tool

Some bee species tend to close every gap and seal every joint in the hive with propolis. The hive tool is a handy piece of metal used to separate boxes, scrape off odd bits of beeswax and separate frame-ends from the supports. It is possible to use an old knife for this job, but knife blades tend to be too flexible and give insufficient leverage. Village blacksmiths can easily produce suitable implements.

Harvesting and processing

Honey is harvested at the end of a flowering season. In traditional or top-bar hives, the beekeeper selects combs which contain ripe honey covered with a fine layer of white beeswax, usually those nearest the outside of the nest. Combs containing pollen or developing bees are left undisturbed. Honeycomb can be simply cut into pieces and sold for a premium price as fresh cut-comb honey. Alternatively, the honeycomb can be broken and strained through muslin or another form of filter to separate the honey from the beeswax. After the honey has been separated from the combs, the beeswax can be melted gently over water into a block. Honeycombs harvested by honey hunters can be processed in the same way.

Honey is obtained from frame hives by spinning the frames in a centrifugal extractor. The empty honeycombs are then returned to the hive. Because the combs are recycled, bees put effort into honey production rather than beeswax comb production. This explains why the beeswax yield from frame-hive beekeeping is low compared to traditional beekeeping methods.

Cross-pollination

Cross-pollination occurs when an insect moves pollen from one plant to another. It is needed when plant sexes are segregated on different plants, such as melons, or by different periods of flowering of the same plant, such as avocado. Many varieties of fruit trees need cross-pollination; they should be planted so that pollinizer trees are near the main crop trees. Production of hybrid seed crops on a commercial scale creates a special need for cross-pollination by insects: a large population of pollinating insects is needed to carry pollen from rows of male plants to rows of female plants.

Pollination affects crop quality and quantity

Crops vary in the extent to which they benefit from insect cross-pollination. Some crops, such as field beans and mangoes, are self-pollinating but give better yields if pollinated by insects. Many, such as passion fruit, cowpea, sesame, litchi, mustard and cashew, give a substantially increased yield when pollinated by insects. Others such as sunflowers, clover, beans, almonds and melons are completely dependent on pollination by insects and otherwise will not produce crops. Adequate insect pollination affects both the quantity and quality of crops: uneven, small fruit often indicate insufficient pollination. Adequate pollination by insects also ensures that early flowers set seed. This results in a uniform and early harvest and gives the crop the maximum length of time to ripen. Pollination can be as important in crop production as water or fertilizer. With the use of improved cultivars and irrigation, pollination can be the limiting factor. The pollination requirements of all major temperate zone crops are well known. In countries with highly mechanized agriculture, the use of bees for pollination increased greatly during the twentieth century and became an integral part of crop production. Less research has been done on the pollination requirements of crops grown in the tropics.

Protecting pollinators

Many other species of bees and pollinating insects living in the wild are highly important for pollination, in addition to honeybees living in the wild or managed in hives by beekeepers. Many factors have caused a decline in the numbers of these insects available for crop pollination. The most serious threat to pollinating insects is the use of insecticides. Herbicides, grazing or cutting of roadside verges and other destruction of flowering plants remove food sources for pollinating insects. Intensive land cultivation and destruction of hedges, banks and rough verges further reduces the habitat for nesting and hibernation sites for bees. It is in everyone's interest to maintain strong populations of honeybees and other pollinating insects. This means increasing people's awareness of the value of insect pollination, stopping unnecessary pesticide use and increasing forage for bees by including nectar-bearing bushes and trees in planting schemes. Farmers can contribute to the protection of honeybees and their habitats as follows.

- Select and use insecticides with great care; if wild pollinating insects are destroyed, there is a risk of decreased crop yields in the future.
- Never use insecticides when flowers are open; foraging insects work on open blossoms and are killed by sprays at this time; if insecticides must be used, spray at a time of day when crop flowers are closed.

More pollinating insects needed

Paradoxically, intensive agricultural practices diminish the numbers of wild pollinators and at the same time increase the need for them. Larger fields increase the need for pollination while a crop is flowering, yet decrease the ability of the local insect population to pollinate adequately. The tendency to concentrate particular crops in certain areas intensifies this situation, because when the major crop is not in flower there may be insufficient forage from other sources.

In temperate countries, large-scale monocultures have increased the need for pollination, yet have decreased the populations of wild pollinators. A similar dilemma is arising in the tropics, where there has been an increase in mechanized farming and an accompanying increase in field size. In the tropics, however, crop flowering is more prolonged and less intensive than in temperate regions. Where growing conditions are favourable, the same crop species may occur in a sequence of growth stages. Many fruit trees flower and fruit throughout the year, albeit more abundantly at certain periods, and therefore forage for bees may be present at all times.

Increased monoculture in the tropics means that flowering will be more concentrated, so large pollinator populations will be needed for shorter periods. Although pollen sources that allow cross-pollination are naturally present in small mixed farms, special provisions for crop pollination are necessary for large areas of a uniform crop (Free, 1999).

4

Apiculture Theory

Apiculture is the art of beekeeping, but more than just keeping hives and harvesting honey, it includes setting up properly located and constructed hives, making sure bees have access to plentiful sources of nectar, and preparing the harvested liquid honey after it has been taken from the hive.

It also includes being aware of local laws governing the keeping of bees and processing of their products and even marketing the final products. Colonies of bees are highly socialized groups of insects that create their own ecosystems in and around their hives.

A detailed understanding of how this society functions is necessary to get the most out of raising them. Before even beginning to set up a beekeeping hobby or business, gaining a thorough knowledge of bee anatomy, behavior, and ecology can build an invaluable basis for the business. An individual who undertakes the responsibility of keeping a hive is taking on an entire civilization of creatures.

Apiculture does not have to be a large-scale project. Keeping a few hives can give families access to endless honey as well as related products like beeswax candles and fresh honeycombs.

It can also go hand-in-hand with other activities, such as gardening, as a few properly managed hives can provide invaluable pollination services to flowers, fruit trees, and vegetables.

Fig. 1. A beekeeper must make sure that bees have access to sources of nectar

Some of the hands-on processes of apiculture include constructing the hives and upkeep on these outdoor structures that can become weathered by the elements. Handling the bees can be one of the most difficult tasks. A smoker and proper clothing can help the beekeeper keep from getting stung while removing honeycombs or carrying out the delicate procedures of replacing the hive's queen bee.

While it may seem strange to relate management techniques to bees, it is a vital skill. Beekeepers must know what the bees require before they can build a successful hive, including providing consistent sources of fresh water, nectar, and pollen. An individual who is both aware and respectful of the natural cycle of bees will find handling them much easier, and he or she will also be alert to pests and threats to the hive before they become a real problem.

Keeping an eye on rainfall charts, temperatures, climate changes, and information on area vegetation can help the alert beekeeper avoid or prepare for potential problems.

Fig. 2. Keeping bee hives may give families access to fresh honeycombs.

An important part of keeping bees is knowing regional and national regulations regarding beekeeping. Honey is a food product, and as such, its preparation is subject to regulations. There are also rules about where hives can be placed in residential areas, and insurance concerns to be investigated as an ongoing part of the business.

Priority measures

The national apiculture programmes are drawn up every three years.

The priority measures that may be included in the programmes are:

- **Technical assistance to beekeepers and groupings of beekeepers**

Technical assistance is designed to enhance the efficiency of production and marketing by introducing better techniques. Courses and other training measures for beekeepers and those responsible for groupings or cooperatives relate in particular to such areas as breeding and disease prevention, collection and packaging, storage and transport of honey and marketing.

- **Control of varroasis**

The aim of varroasis prevention is to reduce expenditure incurred in treating hives or to ensure that part of the cost is met. This disease is caused by a parasite and, at the present time, is severely diminishing the honey yield in the EU, leading to the loss of bee colonies if not treated. As varroasis cannot be completely eradicated, the only way to avoid its consequences is to treat hives with approved products (those that do not leave any residue in the honey). A financial contribution was also considered necessary to ensure that beekeepers do not treat hives with unauthorised chemicals or ineffective remedies.

- **Rationalisation of transhumance**

  Aid for the rationalisation of transhumance is intended to assist with managing the movement of hives in the Community and with providing locations for the strong concentration of beekeepers during the flowering season. Transhumance management can be facilitated by such measures as a transhumance register, investment in equipment and maps of honey varieties.

- **Measures to support laboratories carrying out analyses of the physico-chemical properties of honey**

  Support measures for carrying out analyses of honey are intended to improve the marketing of honey. Financing of analyses of the physico-chemical properties of honey according to its botanical origin provides beekeepers with precise knowledge of the quality of the honey harvested, and enables them to get a higher price for their product.

- **Measures to support the restocking of hives in the Community**

  The restocking of hives makes it possible to compensate for losses of bees, and therefore of production, by funding activities to promote queen production or purchasing of bee colonies.

- **Cooperation with specialised bodies for the implementation of applied research programmes in the field of beekeeping and apiculture products**

  The opportunity provided by the Regulation for including specific applied research projects for improving honey quality in the honey programmes, and dissemination of the results of such projects, can help to increase producer incomes in particular regions.

Notification and monitoring

In order to access the direct aid financed by the EAGF budget for apiculture and to support these measures, Member States must notify their tri-annual apiculture programme to the Commission by 15 April of the first year of the three years period applied by the programme.

The national programmes shall:

- Describe the situation in the sector;
- State the aim of the programme;
- Describe the measures and specify the costs of implementing them;
- State the estimated costs and financing plan broken down yearly at national and regional level;
- Mention the applicable laws, regulations and administrative provisions;
- State the organisations and bee-keeping cooperatives that the Member State collaborated with in drawing up the programme;
- Set up the monitoring and assessment arrangements of the apiculture programme.

Besides the apiculture programmes, Member States must also forward to the Commission documents specifying how the programmes are to be monitored in order to verify compliance with the terms on which the national aid is granted.

Implementation reports

According to Article 184(2) of the Single CMO Regulation Commission presents to the European Parliament and to the Council, a report on the implementation of the apiculture programmes in Member States every three years.

Caring for Extracting Combs

Good combs for producing extracted honey can be reused for many years. As a result, a given number of colonies can produce more extracted honey than comb honey because they are not held back by the need to make new comb during the nectar flow.

After combs are extracted, they still contain some honey and are usually referred to as "wet" combs. Beekeepers do not agree on how such combs should be handled except that they must be fumigated. The combs can be returned to the colonies to be "dried, " then taken off, fumigated, and stored. This involves a lot of work, especially when the bees cluster in the supers rather than going back down into the lower hive bodies. Placing the supers above an open inner cover will not always prevent such behavior.

It is best to store the combs with the honey on them. The bees move into them quickly when they are placed on hives the following season, and the bees benefit slightly from the extra honey. The bees clean out and liquefy any granulated honey so that such combs will not induce granulation of the new crop any more than will combs freed of honey in the fall. If you prefer to clean the combs, do not do so by exposing them in or near the apiary.

This may induce serious robbing that could damage your colonies and spread disease. In addition, the combs may be damaged by the frenzied activity of the bees as they clean them out.

CONFINING BEES

Bees can be confined to their hives for short periods to move them, to protect them from pesticides, or to keep them from bothering people or animals nearby. Whatever the method or material used to keep them from leaving the hive, action must be taken when the bees are not flying, either during the night or in cool or wet weather.

The simplest closure is a V-shaped piece of window screen or hardware cloth pushed into the hive entrance (Fig. 49). Any other openings must also be screened or closed at the same time. This method of closing hives is suitable only for very short periods when the weather is not hot. With stronger colonies, or during hot weather, or for longer periods, the colony needs extra space in which to cluster.

This can be provided by using an entrance screen and a top screen. These screens have wooden frames that give the bees space in which to cluster outside the hive (Fig. 50). A shallow super with one screened surface makes a good top screen that can be stapled or cleated to the hive.

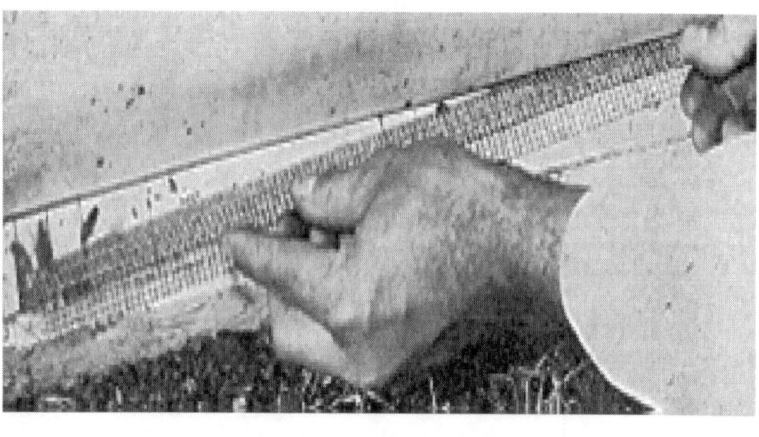

Fig. 3. Closing a hive with a V-shapeed piece of 8-mesh hardwire cloth

Fig. 4. A hive with top and enterance screen in place for moving. Bees can move into both screens to cluster and to ventilate the hive

Bees can also be confined by covering the hives with plastic sheeting, burlap, or other materials. The coverings are draped loosely over the hives and held down by soil around the edges. Black plastic sheeting is suitable for only a short period early in the day because it heats up rapidly in the sun. Burlap can be used to keep bees confined for a day or more. In hot weather it can be kept wet to cool the bees beneath it.

Dividing Colonies

Splitting a strong colony of bees into two or more separate units is an important technique in beekeeping. It provides new colonies to replace losses or to increase numbers of colonies. It is also a method of swarm control, and can be used to make up small colonies (nuclei) for rearing or holding queens. To divide a colony you must first find the queen. If you are unable to find her in a large colony, put a queen excluder between the brood chambers and close the hive. Three or more days later examine the colony again. The queen will be in the brood chamber that has combs with eggs. She is easier to find in a single hive body.

Colonies may be divided initially within the same hive by using a double division screen. Place the old queen with about half the combs of brood, mostly unsealed if possible, in the bottom brood chamber. Add an extra hive body with empty combs or combs with some honey if it is needed. Put the double division screen on top of the second body with the entrance facing the rear of the hive. Above it put the second brood chamber containing five or six frames of brood, mostly sealed, and two combs of pollen and honey on each side. This hive body initially should contain about twothirds of the bees. You must shake many extra bees into it from the combs of the bottom chamber (Fig. 51) because the older field bees will return to the bottom story leaving only the younger bees in the new colony on top. The new division may be too weak to keep the brood warm if an insufficient number of bees is present. A caged queen should be introduced into the top colony within 2 hours for best results but no later than 24 hours after making the division. After the queen is accepted and laying well, the new colony can be put on a bottom board within the same apiary. Fewer bees will be lost, however, if it is moved at night to a new location 2 or more miles away.

Fig. 5. Shaking bees froma comb into the hive. One or two sharp shakes remove most bees with little antagonism if the bees are smoked first.

Divisions can also be made directly into a complete second hive. In this case, give the new colony more than half the bees and four to six frames of sealed brood. The hive may be placed near the parent colony.

However, it is better to screen the entrance of the new hive while making up the colony and then to move it to another location at least 2 miles away to prevent bees from returning to, the original colony. Put the screened colony in the shade after you finish the division so that it will not be damaged by overheating. As soon as it is moved to the new location, smoke the entrance and take out the entrance screen.

The same general system of dividing can be used to make small nucleus colonies. For a three-frame nucleus, take one or two frames of brood and bees and a frame of honey from a strong colony. Pick mostly sealed or emerging brood that fills only a third or one-half the frame if possible. Before you put all the combs into the hive, shake two or three additional frames of bees into it. Introduce a queen or a queen cell as soon as possible but not later than 24 hours after making the nucleus. Although the nucleus can be left in the home apiary, it will do better if it is moved to another location.

New colonies of all sizes may be made from brood, bees, and combs from several colonies. Use the same general techniques as explained above and assemble the colony with sufficient bees and stored honey and pollen to get it started. In making divides and nuclei use small- to medium-sized brood patterns in preference to very large areas of brood. The new colony may not be able to care for a large amount of brood. By using sealed brood, you reduce the number of bees in the parent colony and rapidly increase the number in the new colony.

Feeding Bees

Honey and sugar. More honey bee colonies die from lack of honey than from any other cause. To prevent such losses the beekeeper must know the two periods when the colonies need additional food and the best way to give it to them. The late winter-early spring period, after brood rearing begins, is the most critical period. Feeding may also be needed in the fall if the summer nectar flow was a failure or if too much honey was taken from the hive for home use or for sale.

A comb of honey put into the hive beside the brood nest is the simplest feeder. Combs from hives with a surplus can be given to colonies short of food, so long as American foulbrood disease is not present. Extracted honey can be fed as a syrup by diluting it one-fourth to one-half with warm water; however, do not feed bees any purchased honey. The addition of Terramycin will not make such honey safe for your bees because the drug does not kill the spores of American foulbrood, and these spores can cause infection long after the drug has become inactive.

Feeding excites and stimulates the colonies being fed. This excitement can set off robbing in the apiary, particularly of those colonies receiving the food. To reduce or prevent robbing, and the possible loss of colonies, feed bees late in the day after most flight activity has ceased. Unless the weather is very warm and all colonies are strong, reduce the width of hive entrances with cleats so that colonies can better protect their entrances if robbing takes place.

Table sugar, either beet or cane, can be used in place of honey to feed bees. They will accept sugar in several different forms, including syrup, fondant, and dry granules. Any solid form of sugar must be liquefied by the bees before use and then reduced to the consistency of honey if it is stored. It takes moisture, sometimes body fluids, to liquefy fondant and dry sugar. Bees must also expend energy to remove water from syrup.

It is estimated that they use 4 to 5 ounces of sugar to eliminate each extra pound of water in the syrup. If you want to provide the most stored food in relation to the amount of sugar fed, use a concentrated sugar syrup made from 16 pounds of sugar to each gallon of hot water. At this rate, 100 pounds of sugar will make 13 or 14 gallons of syrup. When fed in the fall, this concentrated syrup will provide up to 11 pounds of stores for each 10 pounds of sugar fed. The process is less efficient when bees are rearing brood in tile spring. You can make a similar syrup by using two volumes of sugar to one volume of hot water.

Dry, granulated sugar can be used for emergency feeding but is not suitable for colonies that must have food immediately to survive. Place it within the hive in an open container above the frames or on top of an inner cover around the open center hole. Brown sugar, molasses, plain corn syrup, and other similar materials containing sugar should not be used as feed, but high fructose corn syrup may be fed safely.

There are several methods and types of equipment used to feed syrup to honey bee colonies. The beginner often uses an entrance feeder that holds a quart jar. It is easy to use but has some serious disadvantages. For package colonies and other small colonies tile syrup in the feeder gets too cold and is too far from the cluster during cool weather. If you use one, put it on the side of the entrance nearest the brood nest and close part of the entrance beside the feeder to reduce the chance of robbing.

Tile division-board feeder hangs inside the hive in place of a frame. It holds about 2 quarts and can be refilled without removing it from the hive. It provides food quickly to strong colonies but is not a good choice for slow, stimulative feeding.

The best all-purpose feeder is the friction top can or similar large containers. Five- and ten-pound honey cans, unused paint calls, and gallon glass or plastic jars can be filled with syrup and inverted above the cluster. The feeder can be set within an empty]live body, either directly on the frames or over tile hole of an inner cover.

Leave part of the inner cover hole exposed so bees can get out. If tile feeder leaks, the bees may collect the syrup and keep it from running outside the hive where it will attract robber bees. For slow feeding and stimulation, punch 5 to 10 holes in the feeder lid with a threepenny nail. For winter or emergency feeding, use 20 to 30 holes. In place of nail holes, you can use 40-mesh metal screening to cover a 2-inch hole in a plastic feeder lid. Heat the screen and press it into the plastic.

Fig. 6. A division-board feeder within a hive body. A wooden float is needed inside the feeder for the bees to stand on when taking syrup.

Fig. 7. A plastic jar in use as a feeder over an inner cover. An empty hive body and the cover enclose the feeder.

There are two emergency methods of feeding colonies quickly. One method requires an open tub or trough to hold sugar syrup and some kind of float on which the bees can land. A temporary cover will give protection from the rain. This is a poor method of feeding: the weaker colonies may not get the food they need to survive, any nearby colonies may take part, and the colonies may rob each other.

A better emergency method makes use of combs filled with heavy sugar syrup. To fill them, use a sprinkling can, a coffee can with the bottom full of nail holes, or a garden sprayer free of insecticide residues. Hold the empty combs over a tub or large pan and sprinkle or spray the syrup into the cells of the comb. With both sides filled, a comb will hold several pounds of syrup. Place two or more filled combs next to the cluster of any colony that needs food.

Pollen, pollen supplements, and substitutes. Pollen for feeding bees is obtained by the use of pollen traps that remove fresh pollen pellets from the legs of incoming field bees.

For only a few colonies, combs can be filled with the pellets and used immediately or stored for later use. For larger numbers of colonies this method is impractical. To fill a comb, pour fresh pellets from a pollen trap into the cells on one side of an empty comb, tap the comb several times to settle the pellets, and put it into a strong colony overnight. The bees will pack the pollen into place and the process can be repeated the next day for the other side of the comb. The pellets from the trap also may be dried or frozen for later use.

Pollen substitutes are protein materials, used alone or in mixtures, that bees can use temporarily for rearing brood. Among them are several kinds of soy flour, brewers' yeast, casein, and dried milk. When the materials are mixed with natural pollen they are called pollen supplements. Bees eat the supplements much more readily than they do the substitutes because they are attracted by the pollen.

Pollen substitutes are available from beekeeping supply companies and from feed companies. You must add your own pollen to make a supplement. Do not use purchased pollen because it may carry bee disease organisms.

These foods can be given to bees as a dry mix in open feeders in the apiary or as a moist cake or patty on top of the frames in the hive directly above the brood nest. For open feeding, a pan or dish of the mixture can be placed in any open-front box with an overhanging cover to keep out rain and dew. Large-mesh chicken wire over the opening lets bees in but keeps out other animals.

There are many different formulas for pollen mixtures; they may be purchased ready to use or mixed as follows:

Dry mix:

2 lb. brewers' yeast

6 lb. soy flour

2 lb. dried, ground pollen pellets, if available

Pollen cake:

15 lb. soy flour, or soy flour-brewers' yeast mixture

5 lb. dried pollen pellets, if available

13 lb. water*

27 lb. Sugar*

* = combined these elements make sugar syrup

Fig. 8.Pollen cake in the hive above the brood nest.

Fig. 9. Bees visiting a box containing dry pollen mix. The lid is hinged for ease of refilling the pan containing the mixture.

Add enough warm water to the pollen pellets to make a slurry, which you mix with the sugar syrup. Gradually add the mixture to the soy flour, kneading it into a dough. Add extra water or soy flour as needed. Put 112 to 3/4 pound of the dough between sheets of waxed raper and flatten to 3/8-inch thickness. If you have no pollen, use 20 pounds of soy flour or a pollen substitute mixture.

Begin feeding the dry mix or pollen cake in February or early March and make it available to the bees continually until natural pollen is available.

Protecting Stored Combs

Honey combs not protected by a strong colony of bees must be treated to prevent damage from the greater wax moth and other moth pests. A beekeeper must assume that any equipment removed from the hives during the bees' active season may be infested. Moth eggs and young larvae are difficult to see. The equipment must be treated to kill all stages of the moth (egg, larva, pupa, and adult). It must also be guarded against later infestation as long as it is in storage.

There are two fumigants approved for killing wax moths; carbon dioxide gas (CO2), and paradichlorobenzene (PDB).

Carbon dioxide is the only fumigant approved for comb honey. It must be used in a relatively airtight room or container in which you can hold a concentration of 98 percentC02 for 4 hours at a temperature of 100 degrees F. (38 degrees C.) and 50 percent relative humidity. Although carbon dioxide is not a toxic gas, it must be used with care as a fumigant. At a concentration far below that needed to kill wax moths it can suffocate and kill humans.

Any chamber used for C02 fumigation should be designed to include an exhaust system capable of removing the gas in a known period of time. Do not enter the chamber until the system has been operated at least that long. Makeshift methods of fumigation with CO2 may also be dangerous if the heavier-than-air gas flows out of the container being used to hold the combs. Carbon dioxide kills all stages of the wax moth and can be used on empty as well as full combs. Unless comb honey is treated to kill wax moths, it may become unfit for sale soon after being harvested if wax moth larvae begin to tunnel it as they grow and look for food. If you have a limited amount of comb honey or access to a large amount of freezer space, you can destroy any wax moth infestation in comb honey by freezing it. Temperatures of 20 degrees F. (- 7 degrees C.) will kill all stages in 4-1/2 hours. At 5 degrees F. (- 1 5 degrees C.) only a 2-hour exposure is needed.

Paradichlorobenzene is a white crystalline material that vaporizes slowly in air. The gas is heavier than air, nonflammable, and nonexplosive. Place approximately 6 tablespoonfuls (3 ounces) of the crystals on a paper beneath the cover of a stack of not more than five full-depth supers. The supers should be tightly stacked, with any holes and large cracks covered with tape. PDB kills adult moths and larvae but not the eggs. It also repels moths and should be kept in the stacks at all times for best results. Do not use PDB on combs containing honey because it makes it toxic and inedible. After being treated with PDB, empty combs should be aired for 24 hours or more. After this time, they are ready for use.

In 1981 a new bacterial insecticide called Certan was approved for treatment of empty honey combs. Certan's unique biological formula contains natural bacterium that, when eaten by wax moth larvae, paralyzes and destroys the digestive tract, resulting in the moth's death. Certan does not affect bees, colony activity, or the taste of honey produced on treated combs. It is nontoxic to humans, pets, wild life and beneficial insects. It is a water- dispersible liquid concentrate, thus dangerous fumigant handling and storage is eliminated. Combs are treated by preparing a mixture of one part Certan to nineteen parts water. The mixture is sprayed on the combs with an applicator. Retreatment is necessary after the combs have been reused. If you have only a few supers of stored combs, you should check them regularly during the warm season for any sign of wax moth and treat with Certan. For larger amounts of comb, it is better to fumigate routinely at about monthly intervals unless each stack is protected by PDB. Without such precautions you may find one or more stacks of valuable combs reduced to worthless webs and debris.

Handling Beeswax

Beeswax is an important byproduct of beekeeping and a valuable: ingredient of cosmetics, candles, polishes, and many specialty items. It is also used in the pure form to make comb foundation.

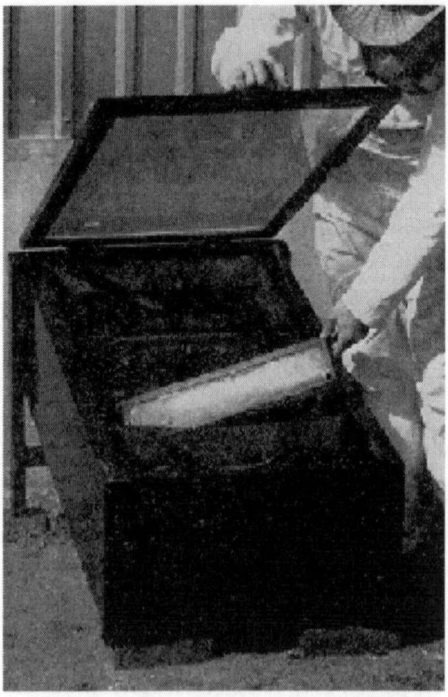

Fig. 10

The beekeeper has several sources of beeswax including cappings from honey combs, damaged combs, and the bits and pieces of comb scraped from hive bodies and frames. From 10 to 12 pounds of wax from cappings is obtained for each 1, 000 pounds of honey, depending on the comb spacing and yield per colony. An additional 1/2 to 3/4 pound per year can be saved from each colony by collecting all the burr combs and scrapings. It is good business to routinely melt very old combs and those with large areas of drone cells, wax-moth damage, and mold. These should be replaced by new frames with foundation to maintain good combs throughout the entire beekeeping operation. A deep super of old combs will yield about 2-1/2 pounds of wax.

Placing a comb in a solar wax melter. The wax pan is removed through the door in front. A screen across the front of the pan for the combs holds back the slumgum while allowing melted wax to run into the lower container.

Wax from cappings is light colored and of a high quality, and should not be mixed with darker wax. Cappings should be melted with a large volume of water in an aluminum, stainless steel, enameled, tinned, or galvanized container. Do not use copper or uncoated steel containers because they discolor the wax. Allow the wax to cool slowly, scrape any impurities from the bottom of the cake, and store it until you have enough to sell.

Large numbers of combs can be rendered in a steam chest or a hot water wax press. The combs can also be taken to a beekeeping supply company for rendering. There is a charge for the service based on the amount of wax recovered. The material called slunigum, which is the residue left when combs are melted in a solar melter or steam chest, is valuable because it contains up to 30 percent wax. It can be commercially rendered for a fee based on the amount of wax secured from the slumgum.

The solar wax melter is a handy piece of equipment for melting comb, cappings, and other sources of wax. It is a sloping pan within a black, insulated box with a glass top, often of double glass (Fig. 56). The heat of the sun melts wax quickly and it runs into a pan where it can be removed in a cake the next morning. The melter can be made any size to fit the needs of the beekeeper. However, it should be relatively shallow and large enough to expose several frames or excluders at a time. It also can be designed to accept cappings baskets made from expanded metal. You can uncap directly into such baskets, allow them to drain, and place them in the melter to render the wax.

Handling Queen Bees

The queen is all-important to the colony, and the techniques of handling and introducing queens are important to success in beekeeping. After learning to find the queen and to evaluate her quality, you must learn to handle her and replace her if necessary.

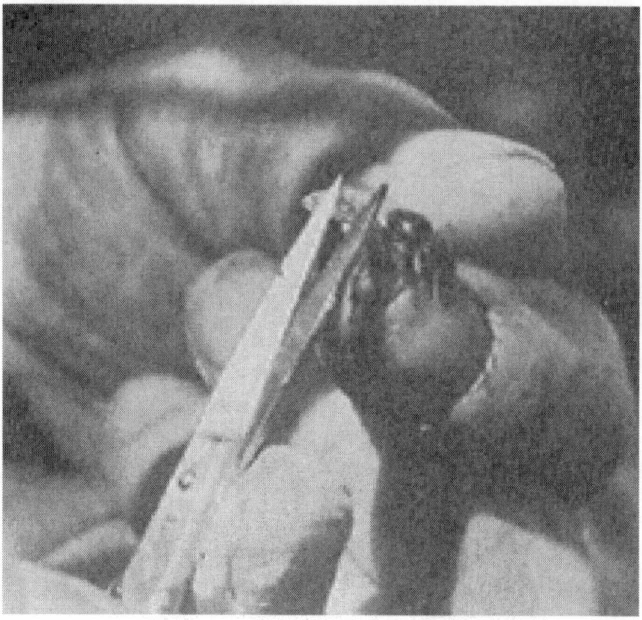

Fig. 11.Holding a queen be in preperation for marking her. Her legs are heald gently but fimly between the thumb and forefinger.

Marking and clipping. The best way to pick up a queen is to grasp both pairs of wings between your thumb and forefinger without pressing her body, especially her abdomen. After getting her up off the comb, hold her against the forefinger of the other hand and trap at least two of her legs with your thumb. Release her wings and you are ready to mark the queen or clip her wings. Before handling a queen, you can practice the technique on drones.

In general, a laying queen cannot fly because of her distended abdomen, and she does not sting when. handled. Mated queens that are not laying, such as those purchased for requeening or in a pack-age of bees, can fly readily and will do so when released from their cages. Handle them only in a closed room or within a screened cage from which they cannot escape, or lightly wet the queen with sugar syrup before opening her cage. Clipped queens cannot fly. Virgin queens fly readily and may also sting occasionally when handled.

Queens are marked to make them easier to find in the hive and to indicate their ages. Queens of the dark-colored races (Caucasian and Carniolan) should always be marked because they are more difficult to find than Italian queens. Fast-drying enamel paint and hot-fuel-proof model airplane dope are satisfactory, inexpensive marking materials that come in a wide range of bright colors. Apply a dot of the material to the queen's thorax, being careful not to get it on her antennae, wings, or membranes. You can practice on drones before attempting to mark a queen. Use a fine brush or, better, a round-headed pin stuck in a cork. Hold the queen briefly after marking her to let the mark dry, and then release her on a comb. In Europe an international marking system of five colors is used to relate the queen's age to her marking.

The colors and years represented are as follows: 1984 - green; 1985 - blue; 1986 - white; 1987 - yellow; 1988 - red; 1989 on - repeat sequence of colors. A German bee-supply company, listed in the section on equipment dealers, sells queen-marking sets with numbered plastic disks in the five different colors. They are of value if you wish to identify each queen individually. The company also sells marking tubes that can be used to hold worker bees for marking.

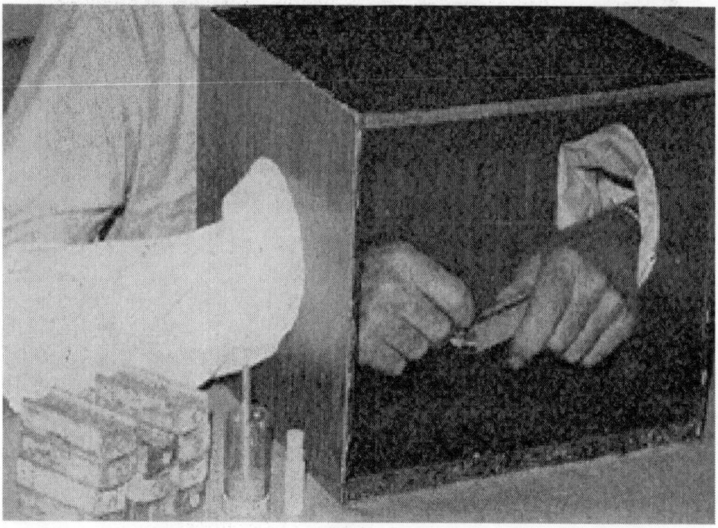

Fig. 12. Using a screened cage to confine a queen and her attendants while handeling them.

Queens are clipped by cutting across one pair of wings to remove about one-third of the longer wing. A fine pair of scissors such as manicure scissors can be used. Clipping was once considered to be a method of swarm control because the first swarm will come back when the queen is unable to fly. It is actually of no help because the beekeeper usually does not see the swarm leave and return.

Shortly thereafter it will leave for good with a virgin queen. Clipping may prevent tile escape of a queen being handled during introduction, but it is more often used as a way to indicate the queen's age. To use it for this purpose, clip the left wing in odd years, the right wing in even years.

Clipping may lead to premature supersedure of the queen, particularly if tile wings are cut so short that the queen's balance is affected as she moves on the vertical combs.

Introducing. Queen introduction is an important part of bee management. A new queen introduced into a mean colony can change its temper in a few weeks and a young queen can more than pay for herself by the increased honey production of her colony. Poor queens should be replaced whenever they are found, and most colonies should be requeened at least every 2 years.

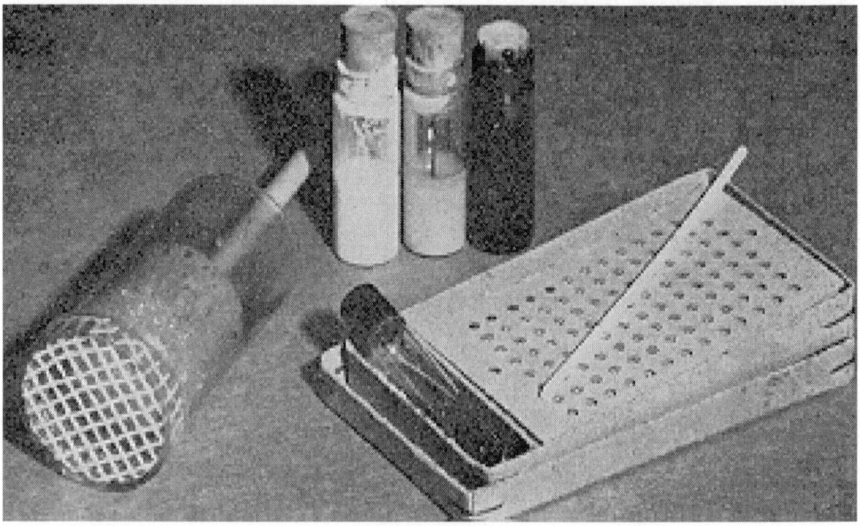

Fig. 13. Bee-marking equipment. Marking disks, in five colors, are used on queens or workers. Worker bees held in the tube are marked through the netting. Model airplane dope in the vials is applied with the head of a pin stuck in the cork.

The first step in replacing a queen is to obtain a young, mated queen from a bee breeder. The queen, together with 6 to 12 attendant bees and a supply of queen-cage candy for food, will arrive in a small wooden cage with a screen top. Give the bees a few drops of water on the screen as soon as the cage arrives.

If you cannot introduce the queen that day, give the bees water twice a day and keep them in a warm place out of the sun. There are holes in each end of the cage that are covered with cork, cardboard, or a piece of metal. In preparation for introducing the queen, remove the cover from the hole on the candy end to expose the candy.

The next step in introduction is to make certain that the colony that is to receive the queen is queenless and without queen cells. Remove and kill the old queen, if there is one, and crush any queen cells with a hive tool to kill the larvae in them. Within 2 hours, place the new, caged queen in the hive. Before that, however, the attendant bees (workers) in the queen cage should be removed.

Many queens are introduced with the attendants present but, because the colony may be antagonistic towards them, the queen will have a better chance of introduction by herself. Remove the cork and let the bees and queen out on a window of a room in which the lights have been turned off.

They will buzz and fan their wings but will rarely sting. As soon as they are all out, pick up the queen and put her, head first, into the hole in the cage. If you don't want to pick her up, hold the cage close to the queen and "herd" her into it with your fingers.

She is then ready to be introduced to the colony just prepared. Wedge the cage, candy end up, between the top bars of two frames in the center of the brood nest. Close the hive and do not disturb it for at least a week.

There are several other types of queen-introducing cages. One of the most useful is the push-in cage shaped like an open-sided box, made of either metal or cardboard. Both kinds work on the same principle, but the metal cage requires the addition of queen-cage candy. Shake the bees off a comb of emerging brood from a colony ready for a new queen. Place the queen beneath the cage on an area with a few cells of honey and emerging bees.

Press the cage at least 1/8 inch into the comb. Replace the comb in the brood nest and leave the hive alone for at least a week. The queen will be released when the bees eat the queencage candy in the tube or tear the cardboard cage to pieces.

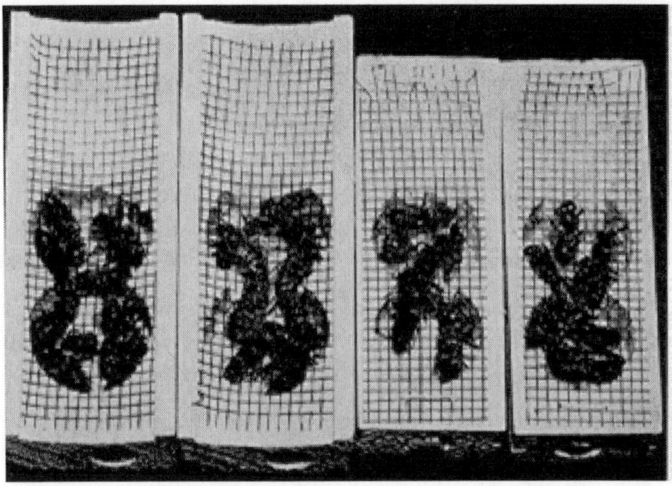

Fig. 14. Queens and attendants in two types of queen cages. One compartment is filled with candy that serves as food for the bees when they are shipped by mail.

Fig. 15. Introducing a caged queen between the combos of a queenless colony.

You can improve your chances of success in introducing a queen if you take into consideration the conditions that favor acceptance. Queens are most readily accepted by small colonies and during a nectar flow. I For this reason, you will have the best results by inroducing queen first to a nucleus or small colony and by feeding the colony a light syrup for several days before and after introducing the queen if there is no nectar available to the colony.

If you do not have a small, queenless colony, prepare one following the directions in the section on dividing colonies. Then introduce the queen to this colony as just explained. When she is laying well, unite her colony with the larger one that needs requeening, after first being sure that it is queenless and without queen cells.

You may unite it by the paper method or by just setting the nucleus in the colony brood chamber after removing an equal number of frames. If you have a large colony that needs a queen immediately, you have no choice but to introduce her directly into that colony. Use the system outlined above for colonies being requeened routinely, when timing is not so critical. If all conditions are favorable for queen acceptance, old and new queens sometimes can be exchanged directly without the use of queen cages.

But this is a risky procedure except under ideal conditions and, therefore, rarely attempted by careful beekeepers. The chance of getting a queen accepted is best if she is held in a cage within the colony long enough to acquire the colony odor and to be fed through the screen by the workers. You can improve the chances of acceptance by replacing the original screen on the queen cage with 8-mesh hardware cloth when you remove the attendant workers.

This larger mesh allows the bees to feed the queen properly, a job that is difficult or impossible through the small mesh of the screen provided on commercial queen cages.

Your percentage of queen acceptance will also be improved if you leave the candy hole of the cage covered for two or three days before removing the cover or cork on a second visit to the colony. Obviously, if you are requeening large numbers of colonies, or if they are located too far away for repeated visits, this procedure is not practical.

In such cases, be sure that there is plenty of candy in the cages and do not punch a hole through the candy to hasten the queen's release. She is much more liable to be killed if released too soon than if she comes out after a delay.

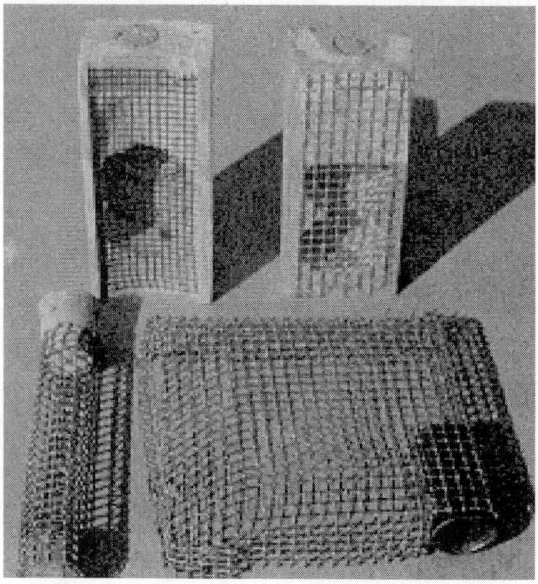

Fig. 16. Different types of queen-introducing cages. The two at the top are commercial, dual-purpose introducing and shipping cages. The bottom two are homemade introducing cages

Fig. 17. Extra queens stored over the super

You can make your own queen cage candy to use for introducing queens. Stir powdered sugar into a small quantity of good-quality honey that you know came from disease-free colonies. You will need about three volumes of sugar to one of honey. When the mixture becomes too thick to stir, knead additional powdered sugar into it with your hands. Form it into a firm ball and let it sit for several hours or overnight. If the ball slumps and becomes softer, add more powdered sugar, then store the finished candy in a sealed plastic bag or other airtight container.

Because honey is a possible source of disease, it may not be used in queen cage candy except for home use. Candy for queens sold locally or shipped must be made with commercial invert sugar, such as Nulomoline or "queen cage syrup, " in place of honey. This invert syrup is mixed with powdered sugar in the same way as described above.

Storing queens. Queens are stored regularly by queen breeders, who must have large numbers available for sale during a short period of time. Beekeepers may also need to store queens when weather is unfavorable for working with bees and when they receive more queens than they can introduce at one time.

The simplest storage method for holding queens as long as one to two weeks is to leave them in their shipping cages together with the attendant workers. When the queens are received, the cages should be separated and placed, screen side up, in a warm place, preferably 85' to 93 degrees F. (290 to 34 degrees C.). However, the cages may be held at a somewhat lower temperature if necessary.

Do not store the cages where they receive direct sunlight. Place a drop or two of water on the screen of each cage every day, but be careful that it does not drip onto the candy within the cage. Cover the cages with a sheet of cardboard except when giving water to the bees.

For longer periods of storage, or when you have no warm place in which to keep them, queens can be held within a honey bee colony. Remove all attendant bees from the cages leaving only the queens. If there is candy in the cage, be sure that the hole on that end of the cage is closed by a cork or by some other material the bees cannot remove easily.

The queens can be placed in a colony with a queen as long as she is beneath an excluder. Place the queen cages, screen side down, over the center frames of the first super above the excluder. Make sure that the bees have access to the screened area where the queen is located in each cage. Naturally the colony must be strong enough so that all hive bodies are well filled with worker bees.

Put a cloth over the cages to hold in the heat and add an empty super before putting on the lid. Larger numbers of queen cages should be placed in a frame modified to hold them within the colony above the excluder.

Put a frame of unsealed brood next to the screened sides of the cages. In strong, queenless colonies, queens may be stored in the center of the brood nest in a frame next to one containing unsealed brood The colony must be given additional, sealed and unsealed brood if the queens are to be stored in it for more than a week. Do not allow the colony to raise a queen of its own until the stored queens are removed In well-maintained, queenless colonies, stored queens can be kept in good condition for periods of one to two months if necessary.

It is better, however, to put them into nucleus colonies of their own for such extended periods. They are available at any time to replace lost or failing queens in large colonies. Unless there is a good nectar flow when queens are being stored in a colony, you should feed the colony with sugar syrup.**Queen rearing.** Queen rearing is one of the most fascinating parts of beekeeping but is beyond the scope of this circular. When you have mastered keeping bees for honey production, try queen rearing. Books on the subject are available at libraries and from beekeeping supply companies.

Hiving Swarms

Swarms are a problem to the beekeeper and to people who are confronted with them in their yards or some other location. The beginning beekeeper can use them to gain additional colonies or to strengthen established ones. However, the time and expense of obtaining them is often more than the small value of the bees themselves. Experienced beekeepers should consider swarm catching a service and charge accordingly for their time and expenses. In some states a license is needed to perform this service.

Swarms are not always gentle and you should wear a veil and use a smoker while working with them. Prepare a single-story hive with nine combs, either empty or partially filled with honey. Foundation is less suitable but can be used if you have no empty combs available. If the swarm is close to the ground, or clustered on a branch that can be cut off, smoke the bees and shake them into the open hive or in front of it. In some cases you may have to shake the bees into a pan, burlap bag, or other container in order to carry them to a hive. If you are successful in getting the queen with the rest of the swarm, the bees will enter the hive and make themselves at home. They should be moved that night to a permanent location. The swarm colony can be allowed to develop, or can be used to strengthen another colony. If you know from which colon), a swarm came, you may put it back after correcting the conditions that caused swarming to develop.

Swarms sometimes come from colonies infected with American foulbrood disease. The honey carried by the bees can infect the brood of the new colony. This serious threat, although not a common occurrence, can be reduced by hiving all swarms on foundation and immediately feeding them 1 gallon of sugar syrup containing Terramycin. Swarms hived on comb can also be fed in the same way, but the protection from disease is less certain. Whether or not you feed the medicated syrup, carefully inspect the colony for disease at least twice before adding another hive body with combs or foundation.

Identifying Apiaries and Equipment

Hives and apiaries located away from the beekeeper's home should be marked to show ownership. Such identification helps to prevent vandalism and theft because it indicates that someone owns the bees. Otherwise people frequently believe that bees have been abandoned because they do not see anyone visit the apiary. Identification is also essential if beekeepers are to be notified of pesticide applications or other farm operations affecting their colonies. One form of identification is the owner's name and address stencilled in large letters on the hives or on a prominent sign beside the apiary. The letters should be at least 1 inch high so that a person who is afraid of the bees can read the sign at a distance.

Frames and other wooden hive parts can be identified by names or symbols stencilled, stamped, or branded on the wood.

Fig. 18. A branding iron and propane torch used to identify frames and other wooden equipment.

Keeping Records

Beekeeping records are of two general types - management and financial. Management records include all the details of the work and observations related to keeping bees. If the information is recorded regularly, it will soon be valuable for planning work, for increasing your knowledge of the biology of honey bees, and for relating management to expenses and income.

Even a simple diary kept up to date can be a worthwhile and enjoyable part of keeping bees. Some of the things to record are local weather data, dates on which nectar and pollen plants bloom, colony losses, colony weight records, and the dates of doing such jobs as spring inspection, supering, removing honey, and extracting.

Financial records are essential for anyone who keeps enough colonies to sell honey. They should be detailed enough to make a financial summary each year for your own information and for computing income taxes and other reports required for business. Farm record books are available from extension service publication offices. Although they are designed for general farming, they can be modified for keeping detailed records of a beekeeping business. Apiary record booklets are also available from several sources in the Midwest. Check with your extension beekeeping specialist or entomologist for details.

Bank-operated recordkeeping services can be adapted for beekeeping enterprises as well as for farm businesses. They simplify recordkeeping for tax purposes and may prove helpful in making short- or long-term loans. Lending institutions need net worth statements and cash flow records in support of loan applications.

Killing Bees

Honey bee colonies are commonly killed when they become infected with American foulbrood disease, when they are living in the walls of a building or in some other unsuitable location, and when they are not going to be allowed to live over the winter. Other management practices also may require that bees be killed.

The insecticide resmethrin is approved for killing diseased colonies of bees whose frames and combs will be destroyed. The insecticide cannot be applied to combs to be reused because treated combs may remain toxic for 7 months or more. Aerosol dispensers containing resmethrin are available from some beekeeping supply companies. Kill the bees when they are not flying, either in the evening or early morning. Close the entrance to confine the bees- the resmethrin causes frenzied activity among the bees when it is applied.

At present, there is no material approved for killing bees as a routine management practice, Cyanide dust and the fumes from burning sulfur have previously been used for this purpose. Because of its extreme toxicity, cyanide is generally not available. The sulfur dioxide gas from burning sulfur is also toxic to humans and does not kill bees as rapidly as does cyanide.

Colonies in buildings should be killed only with an insecticide such as carbaryl, diazinon, malathion, or resmethrin. Fumigants are too dangerous for this purpose. It is important to first locate the brood nest in the wall to learn whether it can be reached by insecticide sprayed or dusted into the flight hole. Sometimes the brood nest is a long distance from the entrance. By tapping and listening you can locate the main group of bees on a cold day or at night when the bees are not flying.

Apply dust or spray at the entrance or through a hole drilled close to the brood nest. Use the material at the concentration recommended on the label for the control of bees and wasps. After the bees have been killed, the dead bees and comb should be removed from the wall and burned or buried. The location will be attractive to other swarms because of the odors present. Filling the cavity with insulation or some other nonflammable material will prevent bees from nesting in the same location again.

Moving Bees

Midwest beekeeping is gradually becoming more migratory as more colonies are moved to sources of nectar and are used for pollination. Even those beekeepers who don't regularly move their hives must some- times move them short or long distances.

The field bees from hives moved short distances -a few feet to as much as a mile or more - tend to return to the original hive location. As they fly out into familiar territory they use the landmarks and flight paths that bring them back to the old hive location. If one hive of a group is moved a short distance, its returning field bees will join hives beside the old location. It is better, if possible, to move all the hives together, a few yards at a time, when relocating them a short distance. Move the bees in the evening or early morning after thoroughly smoking the entrance and any other openings. You may leave the entrance open or screen it closed with a folded piece of window screen or 8- mesh hardware cloth. Careful handling usually makes it unnecessary to fasten the hive parts together to move colonies within an apiary or close to it. However, if you want to fasten them together, do so at least 4 hours before moving the bees.

Most bee moving involves distances great enough to put the field bees into territory unfamiliar to them. No exact minimum distance can be given because it varies with each area and with the foraging distances of the field bees. In some areas a 1-mile move is sufficient, but a good average distance is 2 miles. Naturally, the farther you move the bees the less likely is the chance that some foragers will return to the old location.

The best time to move colonies is about dusk when most of the bees are no longer flying. Early morning is less suitable because the increasing light intensity and rising temperature make the bees eager to leave the hive. If you have difficulties, it is better to have the extra time available at night. A cool, rainy day is also a good time to move bees at any hour so long as the bees are not flying.

The beginning beekeeper who moves bees by truck or trailer should make preparations to complete the job without accidents. Prepare the colonies a day or more ahead of the move by fastening the hive parts together. Use hive staples, lath, or steel or plastic strapping. If you use staples don't put more than four between any two hive parts. Drive them in so they make an angle of about 45 degrees with the crack where the hive parts meet.

Lath cleats are placed on opposite sides of the colony and nailed in place with two or more threepenny or fourpenny nails in each hive part. Be sure to smoke the hive well before you hit it with the hammer. Steel strapping is easy to use and holds the hives tightly but it requires special, fairly expensive equipment.

Plastic tapes are equally good and are easier to fasten with simple equipment. In hot weather, especially with strong colonies, moving screens should be used in place of the regular hive cover. Cover an empty shallow super or similar wooden frame with window screen or 8-mesh hardware cloth and place it, screen side up, over the hive. The bees can cluster in the space and ventilate the colony through the screen. Fasten the hive together with the screen in place.

Cut an entrance screen for each hive the exact length of the entrance and about 4 inches wide. Fold it into a loose V that will slip into the entrance and stay in place. Seal or plug all other holes in the hive.

Fig. 19. Top ventilating screen in place with hives being strapped.

When you are ready to load the hives, put on a veil and light a smoker. Smoke the hive entrance well and wait a minute or two before slipping in the entrance screens. If bees are clustered on the front of the hive you may have to smoke them more than once and wait several minutes before they all go into the hive. Place the hives in a truck or trailer with the entrances facing forward. Arrange the hives as close together as possible in order to reduce bouncing and shifting while en route and tie them in place if possible. At the new location put all the hives in place, smoke the entrances well, and remove the entrance screens immediately. You may remove the top screens at this time or leave them in place with a cover over them until you have time to remove them. The advanced amateur or the commercial beekeeper usually moves bees without entrance or top screens except on occasions when special precautions are needed. Hives moved regularly should have the bottom boards nailed in place and should be equipped with covers that are the same width as the hive bodies. Proper hive equipment and a flat-bed truck with hooks on which to tie the ropes reduce problems in moving bees (Fig. 66). A typical move by a commercial beekeeper may take place as follows.

At dusk the beekeeper drives into the bee yard and prepares to load the hives onto the truck with its headlights off but with the running lights on and engine running. The running lights provide some light to see by, and the vibration of the engine helps to calm the bees after they are loaded onto the truck. With the help of another person, or with a hive loader, the beekeeper quickly places the hives, one to three tiers deep, in rows of five across the truck. Each colony is smoked before it is loaded, and the bees on the truck are smoked peri- odically if they show signs of unrest. As soon as the load is in place, the beekeeper ties each row using a trucker's hitch and a good-quality, 3/8-inch hemp or polypropylene rope.

At the new location, the lights are turned off, the engine is left running, and the smoker is lighted. After the entire load is smoked, the ropes are untied and the hives unloaded. The beekeeper is ready to leave the apiary as soon as the smoker is out and the ropes are coiled. Hive loaders make bee moving a one-man task. They are also useful for handling honey supers and other equipment. Heavy- duty loaders can handle two hives on a pallet or one above the other for loading two tiers at a time.

A tractor with a fork lift can be used for loading pallets with six or more hives. Such palletized hives are preferred by apple growers for use in hilly orchards where a tractor must move the bees to their locations among the trees. The hives are strapped to the pallets and tied with ropes to the truck. In most states, colonies must be inspected and a permit must be obtained before bees can be moved into the state. These procedures may also be required for movement between counties. Before moving your bees, inquire about the regulations at your state Department of Agriculture or other responsible agency.

Fig. 20. Moving bees with an electric hive loader.

Fig. 21. Cradle of hive loader with control buttons. Spring loaded **clamps fit into the hive handholds to support the hive.**

Repelling Bees

When robbing gets started in an apiary, it may be necessary to repel robber bees from weak colonies, open hives, and any equipment stacked in the apiary. The first thing to do is reduce the size of entrances of all weak colonies or nuclei. For extended periods, place a cleat over all but an inch or two of the entrance.

As a temporary measure, stuff grass, leaves, or similar materials into the entrance so that only a small open area is left to be defended by the bees. To make it easier for the bees to remove the material later, do not push it in too tightly. If you must continue to work, expose as little of the hive as possible. Set supers flat on the inverted cover and put wet cloths over the top of them.

Do not set any frames outside the hive or expose honey, syrup, or bits of comb to the robber bees. Under severe robbing conditions cover the open top of the hive with wet cloths, leaving only enough space to examine one comb.

There are no effective repellents available for use on crop plants to reduce insecticide damage to bees. It is also difficult to repel bees from their accustomed watering places such as bird baths and other places where they are not wanted. Solutions containing pine tar, or having the odor of phenol, are slightly repellent to honey bees and may be useful.

Saving Queenless Colonies and Helping Weak Ones

When you are certain that a colony is without a queen, there are several things you can do for it, depending on the type of colony involved and the time of year. The queen lost from a new package colony must be replaced very quickly if the colony is to survive and be productive. Get a queen locally if possible. In some localities, such as the Chicago area, queens are available from bee supply dealers.

Unfortunately, by the time you are sure a package colony is queenless, it may be too late to get either another queen or a replacement package. As a last resort in such cases, you may want to put a swarm into your equipment. By giving your name to the police department you can be notified of the location of swarms.

Even small, queenless colonies can usually produce a queen if they have eggs and young larvae or are given a comb containing them from another colony. Insert a comb with a small amount of brood, less than one-fourth of a frame, into the center of the cluster.

Queens produced under the adverse conditions in small colonies are rarely very good and should be replaced later, but they can keep a colony alive and growing. After a honey bee colony has been queenless for about two weeks, worker bees begin to lay eggs. They do not lay in a neat pattern as the queen does.

They scatter eggs more randomly and put several eggs in each cell. The eggs are usually on the sides of the cell instead of at the base where they are placed by a queen.

The presence of laying workers in a colony makes it difficult to introduce a new, laying queen. The best treatment is to remove the combs in which workers have laid and to replace them with one or more frames of unsealed worker brood with adhering bees.

The added brood suppresses egg laying by the workers, and the young bees are more receptive to a new queen than are the older bees that make up a large part of the population of a queenless hive. As soon as you have added the brood and bees, you can introduce a new queen by placing her cage between the added frames. Do not try to get rid of laying workers by moving the hive or by "shaking out" the combs. Such methods are unsuccessful because laying workers can fly as well as their nonlaying sisters.

Colonies in an apiary are often of different strengths, or populations, especially in the spring. Even those with good queens may be slow in gaining size because of a heavy loss of bees during the winter. You can help the smaller colonies by adding frames of sealed or emerging brood and bees from the larger ones.

Before making such a transfer, find the queen in the larger colony or make sure she is not on any comb being transferred. Add one or more frames with brood and bees, giving one for each four frames covered by bees in the smaller colony. On a nice, warm day when the bees are flying well, you can use another method to help the small colony.

Shake the bees from several frames of brood directly in front of, and close to, the entrance of the weak colony. Select combs from a large colony after locating the queen. The young bees will enter the colony with little or no resistance. The older bees that have previously flown will return to their original colony.

A swarm can also be used to strengthen a weak colony rather than to start a new one. Collect the swarm in a container, such as a cardboard box, from which you can easily dump it. Place an excluder, with an empty deep super above it, over the frames of the weak colony. Smoke both the colony and the swarm and dump the swarm bees into the empty super. Continue to smoke the bees enough so that they move down through the excluder. Find and remove the swarm queen, remove the excluder and super, and replace the hive cover.

These methods of helping weaker colonies do two things for you. They reduce the size of the large colonies and aid in swarm prevention. They also produce colonies of more equal strength that can be manipulated more uniformly.

Honey production can be improved by bringing the colonies into a nectar flow neither weak nor so big that they are ready to swarm. Brood added to package colonies will also help them to reach full strength much faster than colonies not given such help.

Transferring Bees

Many publications have been written about transferring bees from primitive hives, buildings, and trees to modern hives. They usually suggest tearing open the colony and fitting the combs into new frames. Another method uses a screen cone or bee escape over the flight hole so that bees can come out but not reenter the hole. The displaced bees are supposed to enter a hive located beside the entrance.

Transferring bees is no job for the beginner, and it is not worthwhile for the experienced who can obtain all the bees they need by dividing their colonies. Rather than risk the possibility of being seriously stung for little reward, you should resist the temptation to transfer a colony and, instead, should kill the bees or leave them alone. If you want to try removing bees from a building, do the job for a fee, not just for the bees and any honey in the colony.

You might consider transferring bees as a sport or a form of recreation, but it is not a good way to begin beekeeping or to increase your number of colonies.

Trapping Pollen

Trapped pollen is of value for feeding bees. It will become increasingly important as natural sources of pollen become scarcer and as more colonies are used for spring pollination of crops such as apples. A market has developed for pollen for use by commercial beekeepers to feed their colonies.

Pollen traps vary in some features of design but all of the available models have a double screen of 5-mesh hardware cloth that scrapes some of the pollen pellets from the legs of incoming pollen-collecting bees. The pollen falls through another screen into a box or tray where it is inaccessible to the colony and can be removed without disturbing the bees.

The traps remove only part of the incoming pollen and they stimulate colonies to collect more. They probably reduce honey production if used on the same colony for more than a week or two at a time. However, the value of the pollen for supplemental feeding can easily offset the loss of part of the honey crop from a few colonies.

The pollen should be collected from the traps at least three times per week and dried or frozen for storage. If you wish to dry the pollen, use shallow layers exposed to the air or heated at moderate temperatures, not over 140 degrees F. (60 degrees C.), in an oven. Ants, wax-moth larvae, and small beetles are often found in the pollen.

The ants can be discouraged by use of sticky barriers or pans of oil surrounding the supports for the hive. Rain ruins pollen quickly and all traps seem to be vulnerable to it. In selecting a pollen trap, choose the design that will best keep rain out and provide the maximum area of ventilation for the hive.

Fig. 22. Pollen trap and stand that fit beneath the hive. Bees enter through the wide enterance and crawl upward into the hive through the double screen. The pollen falls through the bottom screen and is removed on a tray from the rear of the hive. (Modified from an origonal design by the Ontario Agricultural College in Canada.)

Uniting Bees

Weak colonies are often liabilities instead of assets. This is especially true when they have poor queens or have been queenless so long that laying workers are present. Such colonies will not make any honey and are not good risks for wintering. They should be united with a moderately strong colony with a good queen. Uniting two weak colonies will not produce one strong colony.

Fig. 23. Uniting a small colony with a larger one by the paper method

Kill any queen present in the weak colony and place the hive, without a bottom board, above a single sheet of newspaper over the open top of the stronger colony. Punch a few small slits in the paper to make it easier for the bees to remove it. In hot weather wait until late afternoon so the heat and lack of ventilation will not damage the upper colony. The bees will remove the paper with little fighting and the colonies will be united. Any colonies united in the fall should be checked again before winter to be sure that the clusters are together and that the hive has sufficient stores for winter.

Although the newspaper method is the safest way to unite bees and causes few losses of bees, colonies may be united without the precautions mentioned above. You can unite bees from several hives in the same way as you can make divides and nuclei from frames of brood and bees from several colonies. If none of the queens are of special value, put all the bees together without finding or killing any queens. The youngest queen is most likely to survive and only rarely will all of the queens be killed. The united colony should be checked after a week or two for the presence of the queen and its general condition and arrangement. When colonies are united, the returning field bees from the relocated hives are disoriented briefly. They soon join the united colony and settle down with only minor problems.

5

Apiculture strategies and Planning

Beekeeping is the most widely spread practice in the farming communities of the region and it is an integral part of the smallholder farming system. In the region, the apicultural resources are immense, particularly in the western parts of the region the natural vegetation coverage is relatively high, as a result in this area the honeybee population is dense and production is relatively high. In the eastern parts of the region, in spite of scarcity of natural vegetations, large areas of inaccessible lands for cultivation and livestock grazing are covered with various types of bushes and make this part of region still to remain potential for beekeeping. Besides this, the beekeeping potential of the region is partly attributed to the various cultivated oil crops, pulse and field flowers, which are very important to bee forages and honey production.

Moreover, farmer beekeepers of the region have well developed and long standing traditional beekeeping skills. In the region beekeeping is mostly practiced at a backyard level by keeping beehives either under separate shelters or around the house wall or even inside the house with domestic animals and family members without any problems. In the region beekeepers have relatively better know-how to manage their honeybee colonies.

Moreover, some beekeepers practice migrating their colonies for better forage. However, the level of beekeeping still remains in traditional system and about 94 to 97 percent of bees are still kept in traditional hives with its various limitations (BoARD 2006). But the chances to introduce and adopt improved beekeeping practices are very high since it is based on improving and expanding the existing rich traditional beekeeping practices of the people (BoA 2003).

HONEYBEE RACES FOUND IN THE REGION

Four geographical races of honeybees (*Apis mellifera monticola, Apis mellifera bandasii, Apis mellifera jemenitica* and *Apis mellifera scutellata)* are reported to exist in different ecological zones of the region (Nuru 2002). Among these *Apis mellifera monticola* and *Apis mellifera bandasii* are widely distributed mostly in high and mid altitude parts of the region. Behaviorally, the migratory tendencies of *monticola* and *bandasii* are very low.

Even in the absences of food they remain in their nest up to starving to death. The reproductive swarming tendencies of these bees are also very low. Some colonies reported to remain 5 to 10 years without having reproductive swarm. Compared to others, these bees are relatively gentle, which may be due to the fact that they have been kept very close to human and livestock for many centuries.

The other bee races, *jemenitica* and *scutellata* are found in the western mid and lowland areas of the region. The migratory and reproductive swarming tendencies of these bees are relatively high and are more defensive. Generally the bees of the region are fast in population build up and in exploiting resources in an erratic environment (BoA 2003).

Honey production

Ethiopia has been amongst the principal honey and beeswax producers worldwide for centuries. *Amhara* region is well known for the production of large amount of honey in the country. According to CACC (2003) census report 69, 757. 87 quintals honey is produced annually. This accounts for nearly 25 percent of the total honey production of the country.

In the region many districts from both moisture stress and moist areas are well known for honey production particularly *Gojjam* and *Gonder* are famous in the country for the production of quantity and quality honey. In terms of productivity (yield per colony) North *Gonder*, West *Gojjam*, and *Wag Himra* are reported to be highest.

Moreover, many districts from moisture stress areas are also highly productive (BoA 2003). The honeybee population and annual production and productivity by zone are shown in Table 1.

Table 1. Honeybee population, honey production and productivity of honeybees by zones of Amhara region

Zones	Honeybee colonies	Honey production, kg	Productivity, yield in kg /colony
North *Gonder*	163,457	1,392,384	8.5
South *Gonder*	127,529	897,025	7.0
North *Wollo*	61,736	350,797	5.7
South *Wollo*	118,399	696,306	5.9
North *Shewa*	67,939	344,678	5.1
East *Gojjam*	89,420	644,167	7.2
West *Gojjam*	132,716	1,297,704	9.8
Wag Himra	67,224	659,454	9.8
Awi	67,768	537,304	8.0
Oromia	18,787	155,968	8.3
Total	917,460	6,975,787	7.6 (mean)

Source: CACC 2003; BoA 2003

Socio-economic importance of beekeeping in the regional economy

Honeybees are one of the few assets available to the rural poor. The ownership of honeybees in the region is regarded as an investment and beekeeping has many advantages that help farmer beekeepers to improve their well being. Its advantages can be itemized for the socio economic impact of beekeeping. Successful beekeepers raise their socio economic standing in areas with subsistence agriculture, and farmers in the region can substantially supplement the family income, some times even double it. This means the family is food secured.

Generally, beekeeping is a valuable small scale activity and it plays a significant role in providing nutrition, cash income, pollination service and social benefits to the economy of the region as well as in the subsistence smallholder farmers.

Dietary use

Honey is a useful source of high-carbohydrate food, and usually contains a rich diversity of minor constitutes (minerals, vitamins, and proteins), adding nutritional quality to human diets. In areas where caloric intake is low, the inclusion of honey in the diet will help supply needed carbohydrates. If basic calorie requirements are met, protein foods may be used by the body as protein.

Seasonally it can provide a useful addition to the diet, as it often can be gathered or harvested when other food is limited or monotonous. If harvested and processed properly, honey may be stored for long periods of time with no adverse effect; therefore, it can be use in times of food shortage. As Crane (1990) noted compared with industrialized countries, honey consumption is generally low in developing countries. There is no exception in Ethiopia as well as in Amhara region. This is because honey is regarded more as a medicine or tonic, rather than a daily food. The tradition of using honey making mead (locally known as *'Tej'*, the national drink, is possibly Ethiopian's oldest alcoholic drink) must not be neglected in the dietary discussion (Bradbear 2003).

Some studies conducted in the region have shown that the percentage of honey utilized for household consumption ranged from 23. 0 percent to 36. 5 percent (CACC 2003; Kerealem 2005). The proportion that is eaten is normally taken at harvesting and often said to be the lowest quality part of the harvest that includes: bee-brood, unripe honey and dark color honey.

As a source of income

Beekeeping is one of the most important income-generating activities in the rural communities. The main emphasis given on honey production is as a cash crop. Honey has good domestic market all the year round with slight price change at different market points. In the region honey selling helps for the diversification of the incomes of farmers. Some farmer beekeepers of the region reported to earn up to 3000 *Birr* (about US$ 353) annually from honey selling only which contribute the largest portion of their annul incomes (BoA 2003). These facts indicate the high potential of apiculture as source and means of diversifications of incomes for the rural communities. Many resource poor farmers sell their honey to the local markets and use income to purchase livestock, agricultural inputs, food crops and other household items. Many beekeepers sell their honey mixed with beeswax without further processing. Honeybees can also be sold to meet cash requirements.

Pollination service

Honeybee is also believed to play a significant role in the economy of Ethiopia through pollination services. Pollination is one of the most important factors that affect seed production in agricultural crops. In Ethiopia, an experiment was conducted to evaluate the effect of honeybee pollination on Niger (*Guizotia abyssinica*) and the result revealed that honeybees increased the seed yield of Niger by about 43 percent (Admassu and Nuru 2000). These indicated that honeybees have a vital role in increasing food production and overall agricultural productivity.

Social benefits

Beekeeping also has considerable socio-cultural significance. In the region farmer beekeepers are appreciated by the community for their sweet product. Moreover, production of honey mead (*'Tej'*) as a local festival drink and the use of beeswax for making of votive candles is an integral part of the cultural heritage within the many ethnic and religious groups.

Development attempts in apicultural sub sector

In Ethiopia the extension program on beekeeping activity was started in 1978 together with other activities of Livestock Department of Ministry of Agriculture. In order to start beekeeping extension activities the program prepared was first to make trials of improved hives in areas selected as demonstration sites. *Finoteselam* in *Gojjam* was one of the selected sites used to demonstrate improved beekeeping in the region. Thereafter during the*Dergue* regime many box hives have been distributed mainly to farmer producer cooperatives, but all the efforts failed with the abolition of cooperatives and the collapse of the regime.

Following the establishment of the Regional National States in the country, the *Amhara* Region Agricultural Bureau took over the responsibility of the apicultural sector development. Between the years 1993 to 1996 the Regional Agricultural Bureau attempted to distribute frame hives to different zones of the region. During those periods efforts were made to give technical assistance in transferring of colonies from traditional to box hives and rendering of training services to farmers, development agents and district staff. In those periods the focus was only on frame hives technologies. However, from 1997 to 2001 beekeeping a package was developed and three levels of beekeeping technologies (local, top bar and frame hives) were considered for different income groups. Accordingly 1614 frame hives and 8080 top bars were distributed and 20,

450 local hive colonies included and necessary inputs were provided to package participants (BoA 2003).

Though *Amhara* region has enormous untapped potentials for development of beekeeping until recently, apiculture was not a priority in the research activities and development programs of government. The year 2002 brought sever drought in Ethiopia: a country known for its cyclical droughts, which are often followed by famine. Even though the drought did not lead to serious famine, it did, however, force the regional government to embark on other campaigns and it aimed at minimizing the effects of drought in drier areas.

This program is directed at harvesting rainwater runoff, developing springs and sinking shallow wells. And, based on the amount of water harvested, 'Integrated economical important technology packages' were under implementation. In this regard, the regional government formulates strategies to reduce poverty, to bring sustainable development and to make the people food self sufficient in one way or another by categorizing the region into moist and moisture stress areas.

In moisture stress areas more focus will be on livestock production mainly small ruminates, poultry and beekeeping. Therefore, beekeeping, though marginalized by planners in the development themes for a long time, it is a pursuit that can be used to realize our most idealistic visions of regional development. Beekeeping seems to be fairly well adapted to the semi dry and dry land areas where there are few other sources of income.

From 2003 to 2005 market-oriented commodity packages were developed. In this regard emphasis was given for frame hive production and more than 110, 000 frame hives were manufactured in the region. However, due to high price of the hive and other associated problems the number of hives reached to the end users was about 7. 5 percent (BoARD 2006).

Along the efforts of the government some NGOs like SOS Sahel Ethiopia, Agri-Service Ethiopia, Save the Children Fund (UK), World Vision Ethiopia, and the like are most active in promoting beekeeping development in the region. SOS Sahel did a lot to introduce top-bar hives in *Meket* (North *Wollo*) and in many surrounding districts and found to be successful in fast diffusion of the technology.

Agri-Service Ethiopia also involved in promotion of top-bar hives in *Enebse Sar Midir* district (East *Gojjam*). The successes obtained from these NGOs can be scaled up to other parts of the region. However, from the vastness of the region, the number of farmers involved in beekeeping and great potentiality of beekeeping for sustainable development the efforts made are not adequate enough to bring significant changes in the sub sector.

Recently, in the year 2008, the Government of Ethiopia laid dawn foundation stone at *Lalibela* of *Amhara* region to establish "National Apiculture Museum" for honey and other hive products production, processing and marketing. This project may have as an implication for sustainable eco-tourism development and heritage maintenance in Ethiopia.

Research in apicultural sub sector

Apiculture research in Ethiopia is one of the neglected and untouched fields of agriculture. Until very recently, *Holetta* Bee Research Center is the sole mandated institution undertaking adaptive and applied apicultural research work that would support development (EARO, 2000). Some of research activities that have been conducted include in the aspects of bee forage identification, honeybee management, honey and beeswax quality, honeybee enemies and diseases identification, honeybee races identification, low cost top bar hives and foundation sheet technologies developments. Some of the out puts are disseminated to few beekeepers at few places. However, these information and technologies were found inadequate to promote or improve the productivity of honeybees as expected. The centre is currently, the co-ordinating centre for the national apiculture research program.

Nowadays, modest start on apiculture research has been made in Regional Agricultural Research Institutes. *Amhara* Regional Agricultural Research Institute (ARARI) is one of the institutes that given mandate to undertake apicultural research at regional level. The livestock research strategy plan for 2001 to 2005 was designed to be demand-driven, market-oriented, agro-ecology based system approach in problem solving that draws insights from interaction with farmers. Under ARARI there are eight Agricultural Research Centers, of these nearly half of them given mandate to carry out applied research projects on apiculture, meliponicultue and sericulture. Some works have been conducted in the first one while the last two are the new concept. However, this new orientation of research can be hampered to a number of factors. These include lack of trained man power (researcher) and absent of laboratory facilities in the sub sector. Therefore human resource development and construction of laboratory facilities should be given due attention.

Constraints

The prevailing production constraints in the beekeeping development of the country are complex and to a large extent vary between agro-ecological zones and production systems (EARO, 2000). Variations of production constraints also extend in socio-economic conditions, cultural practices, climate (seasons of the year) and behaviors of the bees (Adjare 190). According to the results of this survey, the main constraints of apiculture sub sector in*Amhara* region are indicated in Table 2.

Table 2. Major constraints of beekeeping in the study districts (2006)

No.	Apiculture constraints	Percent of respondents who experience constraints
1.	Shortage of bee forage	20. 0
2.	Pesticides poisoning	17. 5
3.	Lack of skilled manpower and training institutions	12. 7
4.	Low level of technology used	11. 0
5.	Honeybee pest and diseases	10. 5
6.	Shortage of bee colony	8. 3
7.	Marketing problems	8. 0
8.	Absence of policy in apiculture	7. 0
9.	Other constraints	5. 0
	Total	100. 0

Shortage of bee forage

Shortage of bee forage due to population pressure, lack of land use policy and the high demand for farmlands put pressures on mountainous areas to be used for crop production and livestock grazing. These create deforestation, soil erosion and irreversible ecological degradation. Moreover, burning of undergrowth and destroying of forestland for expansion of farmland could trigger a reduction of honey producing floras and foraging areas.

The elimination of good nectar and pollen producing tree species in many areas makes it difficult to maintain bee colonies without feeding (Kerealem 2005). Based on the results of rural households' socio-economic baseline survey, shortage of bee forage was the major constraint of beekeeping in the Amhara Region (BoFED 2002; 2004).

Pesticides poisoning

The use of chemicals and pesticides for crop pests, weeds, *Tsetse* fly, mosquitoes and household pests control brings in to focus the real possibility of damaging the delicate equilibrium in the colony, as well as the contamination of hive products. Of the various kinds of chemicals only insecticides and herbicides are now major problems to the beekeepers. The chemicals used for crop protection are the main pesticides that kill the bees. Moreover, there are two other circumstances in which bees are killed on plants by chemicals.

These are by insecticides applied to non-crop pests such as mosquitoes and*Tsetse* flies and by herbicides applied to plants on which the bees are foraging. Insecticides have a much more dramatic effect on population of bees, thus, the important contribution made by bees to the production of food and human nourishment is being jeopardized.

On the other hand, herbicides, which are commonly not toxic to bees, destroy many plants that are valuable to bees as source of pollen and nectar. The types of chemicals used include Malathion, Sevin, DDT, 2-4 D and Acetone. As it was seen from the beekeeper point of view, poisoning of honeybees by agrochemical has been increased from time to time. Some beekeepers lost totally their colonies due to agrochemical.

Lack of skilled manpower and training institutions

Beekeeping is one of the disciplines which suffered and is being suffering from the lack of skilled manpower, appropriately skilled trainers, training materials and training institutions in the region. Majority of the beekeepers lack the knowledge of appropriate methods of beekeeping. In the country there is no concerned college or university which can provide diploma or certificate level course in beekeeping. Holetta Bee Research Center is the only institute that provides basic trainings to farmers, technicians, and experts. However, this doesn't meet the ever increasing demand of trained manpower in the region.

Low level of technology used

Some studies indicate that the level of beekeeping in the region still remains in traditional system and about 94 to 97 percent of bees are still kept in local hives with its various limitations. An introduction of improved beekeeping technologies to the rural communities are beyond the buying power of the farmers and not easily available for those who can afford it.

Most of the local beekeepers lack the basic tool that would be needed for private work like bee veil, hand gloves, smoker, chisel, and overall (beekeepers suit).

Honeybee pest and diseases

Ethiopia, as one of the sub-tropical countries, the land is not only favorable to bees, but also for different kinds of honeybee pest and predators that are interacting with the life of honeybees (Desalegn 2001). The existence of pests and predators are nuisances to the honeybees and beekeepers. Pests and predators cause devastating damage on honeybee colonies with in short period of time and even over night. According to Kerealem (2005) ants, honey badger, bee-eater birds, wax moth, spider and beetles were the most harmful pests and predators in order of to decreasing importance. Some studies indicate that Ethiopia appears to be free from various honeybee brood diseases and at the same time at low level of adult bees' diseases incidences. A major category of diseases which cause economic loss comprises amoeba, nosema and chalk brood.

Shortage of bee colony

The majority of the beekeepers interviewed said that one methods of obtaining bees is by capturing a wild colony during the reproductive swarming season. However, wild bees habitats are increasingly being destroyed as a result of expansion of farmland and are often suffer from total destruction of their nest. Recent studies also revealed that in most part of the region acquiring honeybee colonies is a major problem to start bee farms and to expand the existing ones.

Marketing problems

It has been observed that in the region the marketing system of honey has many problems. Most of the local markets are far away from the beekeepers and are inaccessible. Beekeepers travel on foot for several hours to sell their honey. The lack of grading systems does not encourage farmers to produce high quality products, thus, the price of honey changes widely based on the good will of buyers. Gezahegne (2001) discussed the constraints to marketing of honey and beeswax in the country and these include low and discouraging price of honey and beeswax in local markets, lower quality of products, lack of market information, absence of organized market channel, transportation problem, lack of appropriate technologies for collecting, processing, packing and storage of honey to keep its natural quality, lack of government support in promoting market development, and low involvement of private sector. Because of beekeepers have limited knowledge of the preferences of their target market, they do not try to make any changes in the quality of their product. Presentation of quality honey is generally poor. Most honey come to market is un-extracted, unstrained and poorly managed.

Absence of policy in apiculture

The livestock sector in Ethiopia has probably suffered more than the crop sector from inappropriate government policies and the apiculture sub sector is no exception. *Amhara* region doesn't have any development strategy and policy in apiculture.

Others constraints

Others technical constraints in beekeeping activities include poor extension systems (absence of coordination between research, extension and farmers), lack of credit service, shortage of records and up-to- date information, shortage of reading materials regarding to beekeeping, and lack of research stations to address the problems related to apiculture.

Opportunities for beekeeping development

Beekeeping is a sustainable form of agriculture, which is beneficial to the environment and provides economic reasons for the conservation of native habitats and potentially increased yield of food and forage crops. In *Amhara* region beekeeping provides a good example of one activity which has a strong local tradition, where there is a local market, and which is environmentally beneficial. Besides the challenges mentioned in Table 2, the main opportunities for beekeeping development in the region are:

- there is still huge potential (local honeybee races and forages) to increase honey production and to improve the livelihood of the farmer beekeepers;
- the Government has recently put in its agenda the need to develop apiculture as one of the strategies to reduce poverty and to diversify national exports;
- since recently, the users of improved hives and appropriate beekeeping equipment and accessories are increasing;
- the presence of micro finance institutes at grass-root level;

- some NGOs are also giving more attention to the sub sector than ever before as an important intervention areas to support the poor and particularly the women. This will give farmers the opportunity to access improved technologies and capacity building; and
- the establishment of bee products development and marketing cooperatives and union. These institutional changes will provide good opportunity to create increasing demand and competitive market for organic honey. This will in turn result in endogenous technological change and overall development in the sub sector.

Suggested research and development interventions

The major problems of honeybee production in the country can be tackled with the appropriate research and development. Researchers are needed for a number of investigation and strengthening simple and adaptive beekeeping practices that best suit the local condition and changing demands. In order to be effective, research must be targeted at the smallholder farmer.

It must also be systems oriented, practical and adaptable. A careful assessment and analysis of the production environment is required in order to formulate apiculture development strategies that will lead to better use of local resources, improve the living standards of poor farmers and ensure the sustainable development of honeybee production.

Inadequate feed sources (nectar, pollen, and water) due to drought and deforestation is a major limiting factor to honeybee production in Amhara region, particularly during the long dry season. Most of the honey plants flower and provide ample nectar and pollen sources after the main rainy season (September to December). Indigenous honeybee floras of various location should be identified and study on species diversity, species composition as well as bee-plant interaction is required.

Research into providing good quality, year-round nectar and pollen source plants for the honeybees is therefore very important. Re-forestation program and utilization policies for the efficient use of natural resources should be developed at regional and national level. Integration of beekeeping to others development activities such as conservation of natural resources and promote the sowing of multipurpose legumes as a soil conservation measures as well as a fertilizer saver and that retain moisture to the soil may help as best means to green the futures. They also provide nectar and pollen for honeybees.

Some browses can also occupy an important role in honeybees feeding, particularly during the dry season. This is because most browse species are drought resistant. These need to study. Generally, honeybee plants such as *Vernonia* spp., *Echinops* spp., *Acacia* spp., *Acanthus* spp., *Helminthotheca echodes* and *Caylusea abyssinica* are well known for their dry period flowering and serving as subsistence forage to bees in dearth periods (Nuru 2002; Gichora 2003).

The agronomic characteristics of these honey plants need to be explored in order to achieve maximum benefit from them. Selection of honey flora suitable for integrated agro forestry-apiculture program should be undertaken. It was generally accepted that a colony must be left with portion of its honey or otherwise be given suitable substitute foods if it is to survive or remain in the hive.

Thus flour of pulses and barley (*'shiro'*, *'besso'*), milk, sugar syrup and the like could be provided to bee colonies as supplementary feed during the dearth periods. Their feeding value and rates of incorporation need to be determined.

It is in no one's interest for bees to be harmed by agroche-micals. Their pollinating activity is a vital part of food production. However, bees can be harmed at any time by chemical sprays present on any plants where bees might forage, even the flowering weeds along the sides of fields. Many beekeepers lose their honeybee colonies every year due to agrochemicals applications. However, this is not considered as a serious problem by the government and other concerned bodies.

Perhaps it is because of the generally low status of mere insects or the fact that bees are not highly visible livestock that the problem is not more fully appreciated. Since some agrochemicals have some residual effect on bee products, these residuals might be hazardous to human health.

Although it is difficult to completely prevent the effect of chemicals on honeybees, their effect can be reduce by strengthening integrated pest management programs, using insecticides of relatively low toxicity and residual effect for bees and other pollinating insects, not applying insecticides toxic to bees when crops are flowering, use proper methods application, and the like.

In this regard there is a need of timely advice and use of educational programs to beneficiaries, chemicals applicators and beekeepers on how to reduce poisoning by proper selection and application in insecticides.

Techniques of beekeeping management (like moving bees out of hazardous areas, supplementary feeding and various protective measures) that will reduce the harmful effect of exposure to insecticides should be developed and practiced.

Research should focus on the effects of agrochemical application on honeybees and means of minimizing their effects as well as on development of non-chemical methods of insect control. There is a need of establish better liaison between the members of the Ethiopian Beekeepers' Association and concerned officials who set up regulations for applying insecticides.

The regional governments should provide insecticide free "sanctuaries" and crop zones where bees can be isolated from insecticides. The government should be also focused on formulating appropriate policies on proper use of agro chemicals and accountability regarding to application.

Rendering training that covers the overall aspects of beekeeping and working tool (equipment) to training of trainers and beneficiaries should be given due attention. Priority should be given for training of technical and professional beekeepers and providing public education on modern beekeeping. Locally beekeepers would be train and then serve as extension agents in their own village.

Development agents who would be involved in beekeeping must have the training first to enable them adequately provide technical assistance to beneficiaries. The farmers selected for beekeeping development program must have the knowledge about biology of honeybees, equipment and working calendar of the areas.

Moreover, modern beekeeping requires close attention and giving technical assistance to the farmers who had little knowledge in operation techniques. In all of the working areas developing honeybee calendar, continuous supervision and some assistance in hive management would help farmers learn more and improve their working capacity better.

Most of the beekeepers in the region have been using local beekeeping technique that result in low hive products. Much of the honey produced by the beekeeper is of very low quality because it is mixed with wax, pollen and brood. Some of the products are even unknown or unexploited. Management systems need to be improved in order to improve the quality and quantity of hive products.

The beekeepers should use a year-round plan of management favorable to the bee colony. The choice of beekeeping technologies varies across geographical areas because of differences in biophysical and economic conditions of beekeepers. Research should focus on development of appropriate bee equipment (hives, protective, honey and beeswax extractors and containers).

Beekeeping is practiced in most parts of the region, but the type of operation varies greatly with respect to geographical area, farming systems and size of operation. Research into the local husbandry system for honeybees is a priority. The indigenous honeybee colonies exist in large numbers in most parts of the region, and a small increment in productivity per colony will collectively lead to large volume of hive products.

Moreover, at any kind of beekeeping development the use of the necessary tools and appropriate hives is essential for effective result. Effort should be done to evolve the most suitable hives to the beekeepers local condition so that they can understand and operate the hives without badly stung and destroying the colony. Knowledge on how to incorporate new technologies profitably into farm level production strategies will become more important.

Honeybee pests and diseases threaten most parts of the regions and cause high mortality rates and severe economic loss. The needs for effective honeybee health delivery service and appropriate control methods in order to reduce diseases, pests and predators constraint remain very important. In this regard research must focus on investigation and diagnosis of factors that endanger the health of local honeybees in different agro ecology zones andestablishing ways of prevention and control measures.

Shortage of honeybee colonies is one of the constraints mentioned by beekeepers. To alleviate the problem queen rearing technique that can easily adopted by farmers should be practiced. Colony splitting technique is one of the easiest and promising ways of colony multiplication.

Therefore, to implement this activity provides practical training to beekeepers and supports them with nucleus hives and other beekeeping equipment have paramount important.

More attention should be given to market development. Linking production and post production components to efficient market information and extension services, infrastructure and marketing schemes, and establishing standards for quality control of bee products should be the major focus in hive products marketing. Farmer beekeepers should be encouraged to establish a "honey and beeswax producer cooperatives", trained in proper management and processing of locally produced hive products, thus creating better potential to the local and international markets. A central collection and processing center for both honey and beeswax should be established, with numerous local collection sites available to beekeepers.

Much of the honey produced from local hives is mixed with wax, pollen and brood and this procedure has not changed because there is a high demand for the supply for the making of '*Tej*'. Grading the incoming crops and basing payments on their quality will give beekeepers an incentive to try new methods of production. More emphasis should be given to unifying and merging the various beekeeping organizations, honey promotion institutions, honey packer and producers in to a strong regional organization with common goals and objectives.

There is a strong need for an apiculture development policy with appropriate guides and well-defined goals in order to attain a thriving production sector with accelerating and environmentally sustainable growth. Technology packages have to be developed through farming systems research.

A research system that benefits small-scale farmers operating in different farming systems and agro-climatic zones should be strengthened. Creating a system that develops a mechanism for strong links between research and extension services should also be an integral part of the envisaged development strategy and policy. A successful apiculture development strategy requires the formulation of natural resource management plans that complement the wider economic and specific agro-ecosystems objectives.

The strategy will also need to consider the social, cultural, political and institutional elements that affect the management of natural resources. On the policy side, issues relating to land-use, bee resource conservation and development strategies, priorities for apiculture development and research capacity have to be addressed.

To encourage the development further, consideration needs to be a number of factors, including improving access to credits, encouraging involvement of women in beekeeping development, making farmers aware of the usefulness recording and developing of reading materials to farmers regarding to beekeeping.

Again, the implementation of action programs require both technical and institutional support and, equally important, government commitment and unlimited cooperation of all concerned. It is difficult to suggest solutions to our economic problems involved in beekeeping development in the region unless we are interested, concerned and dedicated to willingness on the part of all involved to raise and spend more budget on proven research and to disseminate the finding results.

Conclusion

- Beekeeping is a well established practice in the farming communities of the Amhara region and it plays a significant role as source of additional cash incomes and nutrition for many thousand subsistence farmers. However, in spite of its significant economic contribution and its great potential for sustainable development for the region, the attention given to the sector until recently was not satisfactory. Beekeeping has been left for nature with little attempts to support it with technological packages to improve its production and productivity. Thus, the farmer beekeepers in particular and the region in general are not benefiting from the sector as the high level of the economic potential of the apicultural resources would allow.
- In this paper the constraints to production and strategies to enhance development are discussed. Based on the result of this study, the major challenges were shortage of bee forage, pesticides poisoning, lack of skilled manpower and training institutions, low level of technology used, honeybee pests and diseases, shortage of bee colony, marketing problems and absence of policy in apiculture. Despite the different challenges encountered the sub sector, the opportunities for beekeeping development in the region were the presence of natural resources (bees and forage), the current attention of the government to develop apiculture as one of the strategies to reduce poverty, high demand for hive products, the establishments of beekeepers co-operatives/union, the involvement of NGOs in sub sector and the presence of micro finance institutes at grass-root level.
- Therefore, beekeeping could probably be a profitable activity to undertake in most parts of the region. The ownership pattern of honeybee colonies is widespread and nowadays the previously neglected 'micro-animals' are gradually gaining a respectable place in research and development programs. Thus, developing appropriate policy and beekeeping development strategy that would be applicable to the different production systems and agro-ecological zones of the region and give a clear sense of direction is vital to improve the utilization of the region's apiculture resources with a promising economic future.

6

Apiculture Management

Beekeeping does not attract much attention. It is easy to visit villages and not "see" beekeeping, unless actively looking for it. Beekeeping, however, is crucially important for agricultural well-being; it represents and symbolizes the natural biological interdependence that comes from insects, pollination and production of seed. Useful small-scale efforts to encourage beekeeping interventions can be found throughout the world, helping people to strengthen livelihoods and ensuring maintenance of habitat and biodiversity.

Strengthening livelihoods means helping people to become less vulnerable to poverty. This is achieved by helping them to gain greater access to a range of assets, and supporting their capacity to build these assets into successful livelihood activities. This chapter shows the useful role that beekeeping can play in creating sustainable livelihoods.

People who have limited cash or financial savings often have other assets or strengths - as opposed to needs - that can be mobilized. Chambers and Conway (1992) developed what is now the accepted definition of a livelihood: "Alivelihood comprises the capabilities, assets and activities required for a means of living. A livelihood is sustainable when it can cope with and recover from stresses and shocks and maintain or enhance its capabilities and assets both now and in the future, while not undermining the natural resource base. "

BEEKEEPING ASSETS

Individual livelihoods depend on access to many types of assets which fall into five categories: natural, human, physical, social and financial. To understand this, think about your own livelihood and all the diverse assets you depend upon: your skills, access to transport, equipment, telecommunications and the social networks you have been born into or have created yourself. No individual category of capital assets, such as finance, is sufficient on its own to create a livelihood.

Beekeeping is a useful means of strengthening livelihoods because it uses and creates a range of assets. Successful beekeeping draws upon all categories of capital assets, although financial capital is **not** essential for getting started in productive beekeeping.

Natural capital assets

Beekeeping livelihoods are built upon natural resource stocks: bees, flowering plants and water. Bees collect gums and resins from plants and use plants and trees as habitat for nesting. Bees are a natural resource, and freely available in the wild. Where bees have not been poisoned, damaged or harmed, they will collect wherever they are able, provided the natural conditions include available flowering plants. Wild or cultivated areas, wasteland and even areas where there may be land mines all have value for beekeeping. Beekeeping is possible in arid areas and places where crops or other enterprises have failed; the roots of nectar-bearing trees may still be able to reach the water table far below the surface. This makes beekeeping feasible in marginal conditions, which is important for people who need to restore their livelihoods or create new ones.

Beekeeping fits in well alongside many other livelihood endeavours because it uses the same natural resources as, for example, forestry, agriculture and conservation activities. Beekeeping provides an excellent bonus in addition to other crops rather than instead of them, because only bees are capable of harvesting nectar and pollen. There is no competition with other insects or animals for these resources that otherwise would be inaccessible to people. Beekeeping ensures the continuation of natural assets through pollination of wild and cultivated plants. Flowering plants and bees are interdependent: one cannot exist without the other. As bees visit flowers, they collect food and their pollination activities ensure future generations of food plants, available for future generations of bees and for people too. It is a perfect self-sustaining activity. Pollination is difficult to quantify, but if it could be measured it would be the most economically significant value of beekeeping.

By definition, a livelihood should enhance capabilities "while not undermining the natural resource base" (Chambers and Conway, 1992). Beekeeping goes beyond this, because it actually helps to sustain the natural resource base. Throughout the world, beekeeping has traditionally been part of village agriculture. Now, as farming practices change, it is essential to ensure that beekeeping is retained and encouraged in order to provide continued populations of pollinating insects.

Human assets

Many societies have considerable traditional knowledge and skills concerning bees, honey and related products. The products of beekeeping are often used by women: the important *tej* (honey wine) industry in Ethiopia, for example, is run by women. Elsewhere in Africa, women brew and sell honey beer. These are the types of human assets or skills needed to create livelihoods within a society. Beekeeping projects have sometimes ignored existing knowledge or implied that it was wrong or out of date, which is worse. The best beekeeping projects recognize existing skills and build on them for greater income generation and to ensure sustainability.

Physical assets

Successful beekeeping enterprises require production equipment and infrastructure such as transport, water, energy, roads, communication systems and buildings. There are many ways to manage bees and obtain crops of honey, beeswax or other products. In sustainable beekeeping projects, all equipment must be made and mended locally which, in turn, contributes to the livelihoods of other local people. Beekeeping can add to the livelihoods of many different sectors within a society including village and urban traders, carpenters who make hives and stands, tailors who make veils, clothing and gloves and those who make and sell tools and containers.

Social assets

Social resources such as networks and producer and marketing associations have great significance for beekeeping development. Local associations provide the means for beekeepers to advance their craft, lobby for the protection of bees, organize collective processing for honey and wax, and gain access to markets. Access to networks at a wider level through non-governmental organizations (NGOs) such as Apimondia and Bees *for* Development helps beekeepers to make national and international contacts, find sources of training, markets and new research results, and improve their understanding of the industry.

Financial assets

Although significant financial assets are not essential to initiate beekeeping activities at subsistence level, they are essential for development of beekeeping enterprises. Successful marketing depends on adequate supplies of containers for processing and packaging. Credit is necessary for beekeeping associations to run collection centres and for traders to buy honey and beeswax.

A good beekeeping project will utilize available assets; it will not depend on imported resources or equipment, such as the beeswax foundation used in frame hives. There are situations all over the world where beekeeping can be especially valuable because it remains an activity that is possible for people living in the most difficult circumstances, perhaps isolated by war or sanctions. This is because bees are almost always available in the wild and equipment can be made from whatever materials are at hand.

Beekeeping outcomes

Beekeeping produces a number of quite different outcomes.

- Pollination of flowering plants, both wild and cultivated, is vital for continued life on earth. However, this essential process is difficult to quantify.
- People everywhere like honey, the best-known beekeeping product. Honey is a traditional medicine or food in most societies. Whether sold fresh at village level or in sophisticated packaging, honey generates income and can create livelihoods for several sectors within a society.
- Beeswax is a valuable product of beekeeping, and much of the world's supply comes from developing countries.
- Beekeeping products such as pollen, propolis and royal jelly can be harvested and marketed, although special techniques and equipment are needed for some of these products.
- Beekeepers and other community members can create assets by using honey, beeswax and other products to make secondary products such as candles, skin ointments and beer. Secondary product brings a far better return for the producer than selling the raw commodity. This work strengthens people's livelihoods.
- Products of beekeeping are used for apitherapy in many societies.
- Honey, beeswax and products made from them, such as candles, wine and food items, have cultural value in many societies and may be used in rituals for births, marriages, funerals and religious celebrations.

- Beekeepers are generally respected for their craft. Bees and beekeeping have a wholesome reputation. Images of bees are used as symbols of hard work and industry, often by banks and financial institutions.

FOR HONEY PRODUCTION

The dependent factors for honey production are the duration and timing of the honey flow in a certain area. Duration and timing of a honey flow may vary widely depending on local predominant climates, weather during the honey flow and the nectar sources in the area. Good honey production sites are the far northern latitudes. In the summer, as days grow longer, bees can fly and forage for longer hours increasing the production. Migrating beekeepers also take advantage of local bloom of agricultural plants or wild flowers and trees. In mountainous regions a beekeeper may migrate up the mountain as the spring and summer bloom progresses.

It has been shown, that a larger bee colony will produce relatively more honey. Therefore the early buildup and spring feeding and subsequent prevention of swarming are of high priority. Several different methods such as the Demaree method, Checkerboarding and opening up the brood nest have been advocated to prevent swarming.

Techniques to maximize extracted honey production

Once a good location for an apiary is selected, techniques under the control of a beekeeper for maximizing extracted honey production depend mostly on maximizing the number of foraging bees at the peak time of the honey flow. Techniques may include interrupting brood production right before the main honey flow to free up nurse bees for foraging. A main objective is to prevent swarming.

Techniques to maximize comb honey production

Comb honey production requires many of the same techniques that are required for the production of extracted honey. In addition, the colony must be very strong and have comb building traits. Honeycomb for direct consumption as comb honey is always created the same year it is harvested.

Honey combs may also be harvested by crushing the comb and squeezing out the honey. This is the lowest cost method of producing honey. Keepers of the low-cost top-bar hives use this technique to harvest honey. The technique may also be used for the frames of Langstroth hives.

The so-called cut comb are sections of sealed honey comb that are cut out of the frame. If the cut comb is to be consumed not crushed only the purest beeswax foundation may be used.

Techniques for maximizing Ross rounds and cassette production

- Killion Method
- Juniper Hill Method
- Crowding
- Shock Shook Method

For pollination

see pollination management

Techniques for maximizing agricultural crops pollination see:

- Pollinator decline
- Pesticide toxicity to bees
- Buzz pollination

For queen breeding

Techniques to maximize open mating

Techniques to maximize open mating of virgin queens center around having drones of a desired parentage saturate a queen mating yard.

Techniques to maximize artificial insemination

Artificial insemination of honeybee queens is a process used for very selective breeding of honeybee races. In the open mating of queens the source of drones cannot be fully controlled. In artificial inseminaton the source of drone sperm can be fully controlled and be more predictively selected than in open breeding.

For pollen production

Bee pollen is one of the byproducts of the hive. Pollen collection is usually not the main management objective. Pollen is collected by installing a pollen trap at the entrance of the bee hive. There are varying designs for pollen traps. The pollen trap makes access to the hive harder for the foraging bees. In the process of climbing through the pollen trap wires some pollen is loosened from the bee's pollen basketand falls into a collection container. Varying recommendations describe leaving the pollen trap on for a few days or for more extended periods. Pollen collection works best in an area with various pollen sources throughout the year. Fresh pollen can be frozen or dried. It is used for human consumption or fed back to the colony in early spring to speed up brood production.

For propolis production

Propolis is another byproduct of the bee hive. Certain races of bees are more prone to using propolis. Propolis can be collected on special plastic propolis screens. The tendency of the bees is to use propolis as a glue to seal openings that are too small for a bee to crawl through. A propolis screen is usually put in place of an inner cover. It has small openings that are propolized by the bees. The propolis screen can be frozen which hardens the propolis. Once the propolis is frozen it can be easily knocked off and collected. Bee races that propolize heavily are usually not desirable as it makes other hive manipulation more difficult. There is only a limited market for propolis.

For beeswax production

Beeswax may be a major product or a minor byproduct. The management technique that yields the highest amount of wax per hive is the top-bar hive. During the harvest of the honey from top-bar hives the whole honey comb is removed and crushed to extract the honey.

The commercial honey producers use Langstroth hive frames. The honey extraction process yields beeswax from the uncapping process. The highest quality beeswax is almost white. Lower quality beeswax from older cappings or comb is yellow or brown. Beeswax should be rendered and filtered before it is sold. The least amount of beeswax that can be used as such, is produced in Ross rounds or cassette type comb honey production. Wax and honey are not separated and are consumed together.

Tha ability and tendency to build wax comb differs between the honeybee races. It also differs between colonies. A newly hived swarm produces wax and builds comb very quickly.

For royal jelly production

The production of royal jelly is most dependent on the proper genetics of the queen. Queens and drones are selectively bred to increase the production of royal jelly. A good yield per hive is 5 kg per year.

For apitoxin production

Bee venom (apitoxin) is obtained by stimulating the bee with an electrical current that incite them to sting, releasing a drop of poison onto a glass slide. The crystallized venom can be collected and processed. In order to get 1 gram of dry venom, it is necessary to collect the apitoxin of 10, 000 to 15, 000 bees.

For bee brood production

Bee brood as such is generally not a commercial commodity. However, bee brood is edible, and is used as a food in Asia and Africa.

For the production of nucs

Hive management techniques to multiply colonies use the bees natural tendency to swarm by simulating a swarm. Nucs are bought and sold usually in the spring time. The advantage to packaged bees is that the bees are on established frames with a laying queen and developing brood. A fast developing nuc can be transferred to a full hive box and may produce honey in the same year.

Walk-away split

In a walk-away split, frames with eggs and worker bees are removed and the bees will create a queen cell out of a suitable egg. Once the queen hatches, successfully mates and returns to the hive, the hive will be queenright.

Cut down split

For bee package production

A package of bees is made of a queen and 3 to 5 pounds of bees, typically around 2000 bees. The bees are shipped in a cage clustered around a caged queen. The queen is typically unrelated to the bees, so the cage creates a barrier between the bees and the queen. Packages are usually shipped in the spring from regions of mild winter climates to areas that have more severe winters.

7

A Practical Treatise on the Hive and Honeybee

A beehive is an enclosed structure in which some honey bee species of the subgenus Apis live and raise their young. Natural beehives are naturally occurring structures occupied by honeybee colonies, such as hollowed-out trees, while domesticated honeybees live in man-made beehives, often in anapiary. These man-made structures are typically referred to as "beehives". Several species of Apis live in hives, but only the western honey bee (Apis mellifera) and the eastern honey bee (Apis cerana) are domesticated by humans. A natural beehive is comparable to a bird's nest built with a purpose to protect the dweller. The beehive's internal structure is a densely packed group of hexagonal cells made of beeswax, called a honeycomb. The bees use the cells to store food (honey and pollen) and to house the "brood" (eggs, larvae, and pupae).

Artificial beehives serve several purposes: production of honey, pollination of nearby crops, housing supply bees for apitherapy treatment, as safe housing for bees in an attempt to mitigate the effects of colony collapse disorder, and to keep bees as pets. Artificial hives are commonly transported so that bees can pollinate crops in other areas. A number of patents have been issued for beehive designs.

NATURAL HIVES

Honey bees use caves, rock cavities and hollow trees as natural nesting sites. Members of other subgenera have exposed aerial combs. The nest is composed of multiple honeycombs, parallel to each other, with a relatively uniform bee space. It usually has a single entrance. Western honey beesprefer nest cavities approximately 45 litres in volume and avoid those smaller than 10 or larger than 100 litres. Western honey bees show several nest-site preferences: the height above ground is usually between 1 metre (3. 3 ft) and 5 metres (16 ft), entrance positions tend to face downward, Equatorial-facing entrances are favored, and nest sites over 300 metres (980 ft) from the parent colony are preferred. Bees usually occupy the nests for several years.

The bees often smooth the bark surrounding the hive entrance, and the cavity walls are coated with a thin layer of hardened plant resin (propolis). Honeycombs are attached to the walls along the cavity tops and sides, but small passageways are left along the comb edges. The basic nest architecture for all honeybees is similar: honey is stored in the upper part of the comb; beneath it are rows of pollen-storage cells, worker-brood cells, and drone-brood cells, in that order. The peanut-shapedqueen cells are normally built at the lower edge of the comb.

Ancient artificial hives

Bees were kept in man-made hives in Egypt in antiquity. The walls of the Egyptian sun temple of Nyuserre Ini from the 5th Dynasty, dated earlier than 2422 BC, depict workers blowing smoke into hives as they remove honeycombs. Inscriptions detailing the production of honey are found on the tomb of Pabasa from the 26th Dynasty (c. 650 BC), and describe honey stored in jars, and cylindrical hives.

Archaeologist Amihai Mazar of the Hebrew University of Jerusalem cites 30 intact hives that were discovered in the ruins of the city of Rehov(2, 000 residents in 900 BC, Israelites and Canaanites). This is evidence that an advanced honey industry existed in Israel, approximately 3, 000 years ago. The beehives, made of straw and unbaked clay, were found in orderly rows, with a total of 100 hives, many broken. Ezra Marcus from the University of Haifa said the discovery provided a glimpse of ancient beekeeping seen in texts and ancient art from the Near East. An altar decorated with fertility figurines was found alongside the hives and may indicate religious practices associated with beekeeping. While beekeeping predates these ruins, this is the oldest apiary yet discovered.

Traditional artificial hives

Traditional beehives simply provided an enclosure for the bee colony. Because no internal structures were provided for the bees, the bees created their own honeycomb within the hives. The comb is often cross-attached and cannot be moved without destroying it. This is sometimes called a "fixed-frame" hive to differentiate it from the modern "movable-frame" hives. Harvest generally destroyed the hives, though there were some adaptations using extra top baskets which could be removed when the bees filled them with honey. These were gradually supplanted with box hives of varying dimensions, with or without frames, and finally replaced by newer modern equipment.

Honey from traditional hives was typically extracted by *pressing* – crushing the wax honeycomb to squeeze out the honey. Due to this harvesting, traditional beehives typically provided more beeswax, but far less honey, than a modern hive.

There are three basic styles of traditional beehive; mud hives, clay/tile hives, skeps and bee gums.

Mud and clay hives

Mud hives are still used in Egypt. These are long cylinders made from a mixture of unbaked mud, straw, and dung.

Clay tiles were the customary homes of domesticated bees in the eastern end of the Mediterranean. Long cylinders of baked clay were used in ancient Egypt, the Middle East and to some extent inGreece, Italy and Malta. They sometimes were used singly, but more often stacked in rows to provide some shade, at least for those not on top. Keepers would smoke one end to drive the bees to the other end while they harvested honey.

Skeps

Skeps, which are baskets placed open-end-down, have been used for about 2000 years. Initially they were made from wickerplastered with mud and dung but from the Middle Ages they were made of straw. In northern and western Europe, skeps were made of coils of grass or straw. In its simplest form, there is a single entrance at the bottom of the skep. Again, there is no internal structure provided for the bees and the colony must produce its own honeycomb, which is attached to the inside of the skep.

Skeps have two disadvantages; beekeepers cannot inspect the comb for diseases and pests, and honey removal is not easy – often resulting in the destruction of the entire colony. To get the honey beekeepers either drove the bees out of the skep or, by the use of a bottom extension called an eke or a top extension called a cap, sought to create comb with just honey in it. Quite often the bees were just killed, sometimes using lighted sulfur, to allow the honeycomb to be removed. Skeps could also be squeezed in a vise to extract the honey. As of 1998, most US states prohibited the use of skeps because they can not be inspected for disease and parasites.

Later skep designs included a smaller woven basket (cap) on top over a small hole in the main skep. This cap acted as a crude super, allowing the harvesting of some honey with less destruction of brood and bees. In England such an extension piece consisting of a ring of about 4 or 5 coils of straw placed below a straw beehive to give extra room for brood rearing was called an eke, imp or nadir. An eke was used to give just a bit of extra room, or to "eke" some more space, a nadir is a larger extension used when a full story was needed beneath.

A person who made such woven beehives was called a "skepper", a surname that still exists in western countries. In England the thickness of the coil of straw was controlled using a ring of leather or piece of cows horn called a "girth" and the coils of straw could be sewn together using strips of briar. Likenesses of skeps can be found in paintings, carvings and old manuscripts. The skep is often used on signs as an indication of industry ("the busy bee").

In the late 18th century, more complex skeps appeared which had wooden tops with holes in them over which glass jars were placed. The comb was built in the glass jars which made it commercially attractive.

Bee gums

In the eastern United States, especially in the southeast, sections of hollow trees were used until the 20th century. These were called "gums" because they often were from black gum (Nyssa sylvatica) trees.

Sections of the hollow trees were set upright in "bee yards" or apiaries. Sometimes sticks or crossed sticks were placed under a board cover to give an attachment for the honeycomb. As with skeps, harvest of honey from these destroyed the colony. Often the human bee "robber" would kill the bees before even opening their nest. This was done by inserting a metal container of burning sulfur into the gum.

Natural tree hollows and artificially hollowed tree trunks were widely used in the past by beekeepers in Central Europe. For example, in Poland such a beehive was called a "baræ" and was protected in various ways from unfavorable weather conditions (rain, frost) and predators (woodpeckers, bears). Harvest of honey from these did not destroy the colony, as only a protective piece of wood was removed from the opening and smoke was used to deter the bees for a short time.

Bee gums are still used by beekeepers today, for bee species whose honey output is less than that of the more productive honeybee. Unlike most manmade beehives (which are usable only with Apis mellifera andApis cerana), the bee gum allows housing of many more bee species. The bee gum allows the bees themselves to make their hive (hive structure and cells in which to store their honey and pollen).

In some instances, bee gums are also still used, even with bee species that do produce large quantities of honey (i. e. Apis mellifera nigra). However in these instances, part of the reason why bee gums are then used is that this allows the producers of the honey to distinguish themselves from other honey producers and to ask a higher price for the honey. An example where bee gums are still used is Mont-Lozère, France.

It should be noted however that the length of the bee gums used are rather on the short side; unlike regular bee gums, the bee gums are slightly modified artificially (hollowed out artificially and cut to a specific size).

Modern artificial hives

The earliest recognisably modern designs of beehives arose in the 19th century, though they were perfected from intermediate stages of progress made in the 18th century.

Thus, intermediate stages in hive design were recorded for example by Thomas Wildman in 1768/1770, who described advances over the destructive old skep-based beekeeping so that the bees no longer had to be killed to harvest the honey.

Wildman, for example, fixed a parallel array of wooden bars across the top of a straw hive or skep (with a separate straw top to be fixed on later) "so that there are in all seven bars of deal" [in a 10-inch-diameter (250 mm) hive] "to which the bees fix their combs".

He also described using such hives in a multi-story configuration, foreshadowing the modern use of supers: he described adding (at the proper time) successive straw hives below, and eventually removing the ones above when free of brood and filled with honey, so that the bees could be separately preserved at the harvest for a following season.

Wildman also described a further development, using hives with "sliding frames" for the bees to build their comb, foreshadowing more modern uses of movable-comb hives. Wildman acknowledged the advances in knowledge of bees previously made by Swammerdam, Maraldi, and de Reaumur – he included a lengthy translation of Reaumur's account of the natural history of bees – and he also described the initiatives of others in designing hives for the preservation of bee-life when taking the harvest, citing in particular reports from Brittany dating from the 1750s, due to Comte de la Bourdonnaye.

In 1814, Petro Prokopovych, the founder of commercial beekeeping in the Ukraine, invented one of the very first beehive frames. However, for easy operations in beehives the spaces between elements need to be correct. The correct distance between combs was described in 1845 byJan Dzier¿on as 1½ inches from the center of one top bar to the center of the next one.

In 1848 Dzierzon introduced grooves into the hive's side walls replacing the strips of wood for moving top bars. The grooves were 8 mm × 8 mm (0. 31 in × 0. 31 in), the spacing later termed bee space. The Langstroth hive was the first successful top-opened hive with movable frames. Langstroth hive was however a direct descendant of Dzierzon's hive designs.

There are two basic types of modern hive in common use, the movable-frame "Langstroth hive" (including all the size variants) which has enclosed frames to hold the comb and the movable-comb "top-bar hive", which - as the name implies, has only a top-bar to support the comb. Movable frames and movable combs - as opposed to fixed combs - both allow the apiarist to inspect for diseases and parasites and also allow a beekeeper more easily to split the hive to make new colonies.

Bees occupy the new hive "correctly" only if it already contains frames with some honeycomb or wax plates. If only empty frames are present, bees often build honeycomb that does not follow them and cannot be later removed with the frame.

HIVES OPTIMIZED FOR APIS MELLIFERA, APIS CERANA

Langstroth hives

Named for their inventor, Rev. Lorenzo Langstroth, Langstroth hives are not the only hives of this style, but the most common in North America. Langstroth patented his design in 1852 originally for comb honey production; it has become the standard style hive for many of the world's beekeepers. This class of hives includes other styles, which differ mainly in the size and number of frames used. These include Smith, Segeberger Beute (German), D. E. hive, Frankenbeute (German), Normalmass(German), Langstroth hive, Modified Commercial and Modified Dadant, plus regional variations such as the British Standard National Hive.

Langstroth hives make use of bee space so that frames are neither glued together nor filled with *burr comb*–comb joining adjacent frames.

Langstroth hives use standardized sizes of hive bodies (rectangular boxes without tops or bottoms placed one on top of another) and internal frames to ensure that parts are interchan-geable and that the frames will remain relatively easy to remove, inspect, and replace without killing the bees.

Langstroth hive bodies are rectangular in shape and can be made from a variety of materials that can be stacked to expand the usable space for the bees. Inside the boxes, frames are hung parallel to each other. The minimum size of the hive is dependent on outside air temperature and potential food sources in the winter months. The colder the winter, the larger the hive and food stores need to be. In regions with severe winter weather, a basketball-shaped cluster of bees typically survives in a "double-deep" box. In temperate and equatorial regions, a winter cluster will survive in a single box or in a "nuc" (short for nucleus colony).

Langstroth frames are thin rectangular structures made of wood or plastic and have a wax or plastic foundation on which the bees draw out the comb. The frames hold the beeswax honeycomb formed by the bees. Ten frames side by side will fill the hive body and leave the right amount of bee space between each frame and between the end frames and the hive body.

Langstroth frames are often reinforced with wire, making it possible to spin the honey out of the comb in a centrifuge. As a result, the empty frames and comb can be returned to the beehive for re-filling by the bees. Creating honeycomb involves a significant energy investment, conservatively estimated at 6. 25 kilograms of honey needed to create 1 kilogram of comb in temperate climates. Reusing comb can thus increase the productivity of a beekeeping enterprise.

The modern Langstroth hive consists of the following parts:

- Hive stand: the upper hive components rest on this, providing a landing board for the bees and helping to protect the bottom board from rot and cold transfer.

- Bottom board: this has an entrance for the bees to get into the hive.
- Brood box: the lowest box of the hive and where the queen bee lays her eggs.
- Honey super: usually shorter than the brood box, but is the uppermost box(s), where honey is stored.
- Frames and foundation: wooden or plastic frames with wax or plastic sheets with honeycomb impression where bees build wax honey combs.
- Inner cover: provides separation from an overly hot or cold outer cover and can be used as a shelf for feeding or other purposes.

Commercial hives

Commercial hives have the same cross-sectional dimensions as a National hive (460mm x 460mm), but deeper brood box (267mm/10. 5") and supers. The internal structure of the boxes is also simpler, resulting in wider frames (406mm/16") with shorter handles or lugs. The brood box is picked up using small hand holds cut into the external wall of the hive. Supers have this same feature, which some difficult to use when the super is full of honey. Some beekeepers therefore use National supers on top of a Commercial brood box.

WBC hives

The WBC, invented by and named after William Broughton Carr in 1890, is a double-walled hive with an external housing that splays out towards the bottom of each frame covering a standard box shape hive inside. The WBC is in many respects the 'classic' hive as represented in pictures and paintings, but despite the extra level of insulation for the bees offered by its double-walled design, many beekeepers avoid it, owing to the inconvenience of having to remove the external layer before the hive can be examined.

CDB hives

A Irish hive. In 1894, the British Charles Nash Abbott (1830–1894), advisor to Ireland's Department of Agriculture and Technical Instruction, oversaw the design of a new Congested Districts Board (CDB) hive, in Dublin.

Dartington Long Deep hives

The Dartington Long Deep (DLD) hive takes 14 x 12 inch and can take up to 24 frames. It is possible to have two colonies in the brood box as there is an entrance at either end. It has half-size honey supers, which take 6 frames that are lighter than full supers and are correspondingly easier to lift. The Dartington was originally developed by Robin Dartington so that he could keep bees on his London rooftop.

Beehaus

The Beehaus is a proprietary design for a beehive launched in 2009. It is based on similar principles to a Dartington.

Long Box Hive

The Long Box Hive is a single story hive utilizing fully enclosed frames (per the dimensions of Langstroth hives or deeper by variation,) but is worked horizontally in the manner of Kenya/Tanzanian Top-bar hives. This non-stacked style had higher popularity a century ago in the Southeast United States, but faded from use due to lack of portability. With the recent popularity of horizontal Top-bar hives, the Long Box Hive is gaining renewed but limited utilization. Alternative names "New Idea Hive", "Single Story Hive", "Poppleton Hive", or simply "Long Hive".

Top-bar hives

The top-bar or Kenya-hives were developed as a lower-cost alternative to the standard Langstroth hives and equipment. They are becoming very popular in the US due to their alignment with the organic, treatment-free philosophies of many new beekeeping devotees in the United States. They are also popular, owing to their simplicity and low cost, in developing countries. Top-bar hives have movable comb and make use of the concept of bee space.

The top-bar hive is so named because the bees draw their comb from a top bar, suspended across the top of a cavity, and not inside a full rectangular frame with sides and a bottom bar. The beekeeper does not provide foundation wax (or provides only a small starter piece of foundation) for the bees to build from. The bees build the comb so it hangs down from the top bar. This is in keeping with the way bees build wax in a natural cavity. There is some belief that the use of natural wax in a top-bar hive supports the bees' natural systems in ways that improve their health.

The hive body of a common style of top-bar hive is often shaped as an inverted trapezoid* in order to reduce the tendency of bees to attach the comb to the hive-body walls, though this reasoning has become less popular recently. It may be more likely that the trapezoid shape helps to improve the ratio of the weight of the comb to the amount of attachment at the bar and helps to lessen the likelihood of heavy combs detaching from the top bar when being handled or harvested. Unlike the Langstroth design, *this style of top-bar hive is expanded horizontally, not vertically. The top-bar design is a single, much longer box, with the bars hanging in parallel. This common style is sometimes referred to as a horizontal Top Bar hive, or hTBH. Most horizontal top bar hives in the UK are based on the tandem-follower system designed by Phil Chandler, and free plans are available at http://www. biobees. com

Some top bar hives do stack vertically, much like the Langstroth, National, or other frame hives, although vertical designs use bars with gaps between them, unlike the adjacent bars of the horizontal design. Vertical top bar hives have been largely unknown in the United States until about 2010. This is largely due to the work by Dr David Heaf, who, along with wife Patricia, translated a book by Abbé Emile Warré (~1852) from French and implemented some hives in 2007. These are commonly referred to as Warré Hives, but the abbé named them, "The People's Hive, " for their intended economy and simplicity of construction.

Unlike the Langstroth hive, the honey is usually extracted by crushing and straining rather than centrifuging because a top bar does not have wire reinforcement. There is a laterally oriented extractor available, made by Swienty, but these are not overly popular in the US. Because the bees have to rebuild their comb after honey is harvested, a top-bar hive yields a beeswax harvest in addition to honey.

However, like the Langstroth hive, the bees will store most of their honey separately from the areas where they are raising the brood. For this reason, bees are not killed when harvesting from a top-bar hive, unlike the old practice of harvesting from a skep by driving bees out or killing them over a sulphur pit.

Warré hives

The Warré hive was invented by the abbot Émile Warré, and is also called "ruche populaire" (fr) or "The People's Hive" (en), the Warré hive is a modular and storied design similar to a Langstroth hive. The hive body is made of boxes stacked vertically; however, it uses Top Bars for comb support instead of full frames, as a general rule. The popularity of this hive is growing among 'sustainable-practice' beekeepers.

The Warre hive differs from other stacked hive systems in one fundamental aspect: when the bees need more space as the colony expands, the new box is "nadired". i. e. positioned underneath the existing box(es). This serves the purpose of warmth retention within the brood nest of the hive, considered vital to colony health.

Perone hives

The Perone or Automatic Hive was designed by Oscar Perone, and aims at managing bees in a way more closely analogous to their natural behavior. It is favoured by proponents of 'Natural Beekeeping', who seek to allow more natural bee behaviour and minimise intervention into the hive.

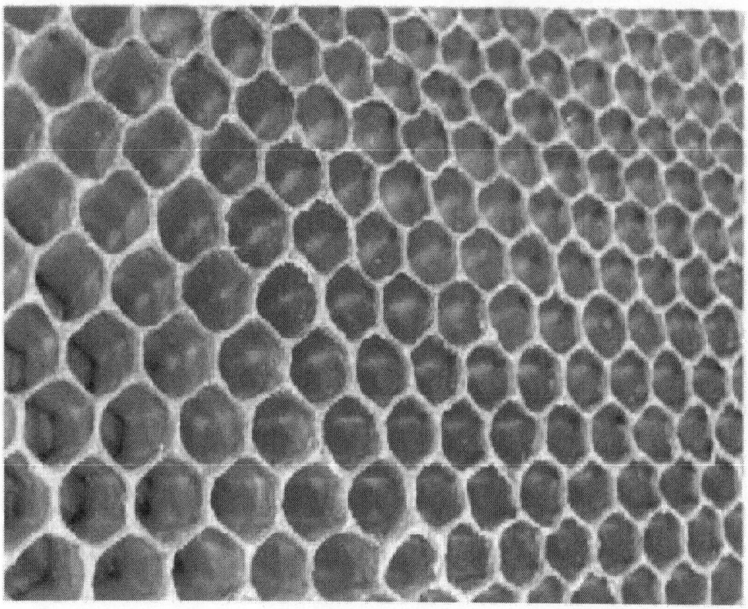

Fig. 1. Bee hive

Perone Hives are very large 2m-high vertical top bar hives that remain the same size all year, split into a bee area underneath, and a bee keepers area above (Mark 1) or side by side (Mark 2). The total hive volume is large, around 280 litres, which it is proposed allows the bees to develop into a 'super-colony' differing in behaviour to colonies in smaller hives. They are managed so that under normal circumstances the bee area is never opened, and the bee keepers area is opened only once per year when any honey is harvested. These hives are managed around bees using their own honey supplies in winter, not replacing these with sugar syrup.

The limitations for direct inspection of this type of hive and its management raise issues around disease monitoring, and they are criticised as sources of varroa for other nearby hives, as the large colony will have many foragers and many standard techniques for managing varroa populations cannot be practiced. However the hive will also have more bees available for defence, and many standard treatments for varroa also cause damage to the bees.

Flow Hives

The Flow Hive is a prototype easy-harvest hive describing a design for supers filled with a split plastic deep comb. The Flow Hive became the subject of a crowdfunding campaign, collecting just over US $12 million as of 2015 Apr 20. It was invented in 2011 and is not derived from a design patented in 1940.

The bees use their wax to seal the splits, build out the comb and fill the cells with honey. According to the 2011 patent, once the comb is filled with honey and capped, turning a crank splits the hexagonal cells in the artificial comb vertically, allowing the honey to flow down the resulting zig-zag channels into a sealed trough and out of the hive. According to the inventors, this allows harvesting the honey with minimal disruption to the bees who are standing on the still capped comb. When drained, the crank is used to reset the comb. The bees then chew open the capping, reseal the splits and refill the cells.

Hives optimized for Meliponines

Hives have been designed for some meliponines such as Melipona beecheeii. Examples of such hives are the Nogueira-Neto hive and the UTOB hive.

Relocation

A common misconception of non-beekeepers is that smoke will cause bees to move their hive. Smoke is used by beekeepers to hide the alarm pheromone emitted by bees during inspections or harvest. One-way traps can be used to reduce the size of a colony, but is a limited time consuming solution. Beekeepers and companies can be hired to do "cut-outs" (a term used to remove bees from a home).

Once honey, pollen and wax are built up in a hive, only physical removal is an option. Some homeowners resort to using spray-foam, but the bees will wait and live off of reserves until the foam is weakened, then will burrow through it.

An important fact in understanding a hive is that less than 10% of the bees will ever leave at once, other than when swarming. A fertilized queen will lay up to 2000 eggs a day. If bees are killed and the hive products are left, heat and time will loosen the wax and honey causing further problems in the future.

Destruction

By other natural organisms

Black bears destroy hives in their quest for honey.

Elephants sometimes also destroy hives, erected by humans as a defense for their crops against elephants. These hives are hung on a single metal wire that encircles the crop field of some farms in elephant territory, in Africa. The installation is called a BeeHive Fence and was conceived by Lucy King.

By humans

Humans often destroy hives of honey-producing bee species in the interest of obtaining honey and other bee products.

Humans may at times also determine that a beehive must be destroyed in the interest of public safety or in the interest of preventing the spread of bee diseases. The U. S. state of Florida destroyed the hives of Africanized honey bees in 1999.

The state of Alaska has issued regulations governing the treatment of diseased beehives via burning followed by burial, fumigation using ethylene oxide or other approved gases, sterilization by treatment with lye, or by scorching. In New Zealand, the treatment of hives infected with the disease American foulbrood with antibiotics is prohibited, and beekeepers are required by law to destroy such colonies and hives with fire.

METHOD OF DESTRUCTION

Spraying the hive with a soap-and-water solution may be effective, since soap dissolves the bees' waxy exterior that protects them from drowning. The procedure is, however, surrounded with cautions, as it angers the bees.

Honey bee

A honey bee (or honeybee) is any member of the genus Apis, primarily distinguished by the production and storage of honey and the construction of perennial, colonial nests from wax. Honey bees are the only extant members of the tribe Apini, all in the genus Apis.

Currently, only seven species of honey bee are recognized, with a total of 44 subspecies, though historically, from six to eleven species have been recognised. Honey bees represent only a small fraction of the roughly 20, 000 known species of bees. Some other types of related bees produce and store honey, but only members of the genus Apis are true honey bees. The study of honey bees is known as apiology.

Origin, systematics and distribution

Honey bees appear to have their center of origin in South andSoutheast Asia (including the Philippines), as all but one (i. e. *Apis mellifera*), of the extant species are native to that region. Notably, living representatives of the earliest lineages to diverge (Apis florea and Apis andreniformis) have their center of origin there.

Fig. 2. Morphology of a female honey bee

The first *Apis* bees appear in the fossil record at the Eocene–Oligocene boundary (23–56 mya), in European deposits. The origin of these prehistoric honey bees does not necessarily indicate Europe as the place of origin of the genus, only that the bees were present in Europe by that time. A few fossil deposits are known from South Asia, the suspected region of honey bee origin, and fewer still have been thoroughly studied.

No *Apis* species existed in the New World during human times before the introduction of *A. mellifera*by Europeans. Only one fossil species is documented from the New World, *Apis nearctica,* known from a single 14 million-year old specimen from Nevada.

The close relatives of modern honey bees – *e. g.* bumblebees and stingless bees – are also social to some degree, and social behavior seems a plesiomorphic trait that predates the origin of the genus. Among the extant members of *Apis,* the more basal species make single, exposed combs, while the more recently evolved species nest in cavities and have multiple combs, which has greatly facilitated their domestication.

Most species have historically been cultured or at least exploited for honey and beeswax by humans indigenous to their native ranges. Only two of these species have been truly domesticated, one (*A. mellifera*) at least since the time of the building of the Egyptian pyramids, and only that species has been moved extensively beyond its native range.

Today's honey bees constitute three clades.

Genetics

The chromosome counts of female bees for the three clades are: *Micrapis* 2N = 16, *Megapis* 2N = 16, *Apis* 2N = 32. Drones of all species have 1N chromosome counts. The genome of *Apis* has been mapped.

Drones (males) are produced from unfertilized eggs, so represent only the DNA of the queen that laid the eggs, i. e. have only a mother. Workers and queens (both female) result from fertilized eggs, so have both a mother and a father. A modified form of parthenogenesiscontrols sex differentiation. The sex allele is polymorphic, and so long as two different variants are present, a female bee results. If both sex alleles are identical, diploid drones are produced. Honey bees detect and destroy diploid drones after the eggs hatch.

Queens typically mate with multiple drones on more than one mating flight. Once mated, they lay eggs and fertilize them as needed from sperm stored in the spermatheca. Since the number of sex alleles is limited – about 18 are known in *Apis* – a queen will most likely mate with one or more drones having sex alleles identical with one of the sex alleles in the queen. The queen, then, typically produces a percentage of diploid drone eggs.

Micrapis

Apis florea and *Apis andreniformis* are small honey bees of southern and southeastern Asia. They make very small, exposed nests in trees and shrubs. Their stings are often incapable of penetrating human skin, so the hive and swarms can be handled with minimal protection. They occur largely sympatrically, though they are very distinct evolutionarily and are probably the result of allopatric speciation, their distribution later converging.

Given that *A. florea* is more widely distributed and *A. andreniformis* is considerably more aggressive, honey is, if at all, usually harvested from the former only. They are the most ancient extant lineage of honey bees, maybe diverging in the Bartonian (some 40 million years ago or slightly later) from the other lineages, but do not seem to have diverged from each other a long time before theNeogene.

Megapis

One species is recognized in the subgenus *Megapis*. It usually builds single or a few exposed combs on high tree limbs, on cliffs, and sometimes on buildings. They can be very fierce. Periodically robbed of their honey by human "honey hunters", colonies are easily capable of stinging a human being to death if provoked.

- *Apis dorsata*, the giant honey bee, is native and widespread across most of South and Southeast Asia.
- *Apis dorsata binghami*, the Indonesian honey bee, is classified as the Indonesian subspecies of the giant honey bee or a distinct species; in the latter case, *A. d. breviligula* and/or other lineages would probably also have to be considered species.
- *Apis dorsata laboriosa*, the Himalayan honey bee, was initially described as a distinct species. Later, it was included in *A. dorsata* as a subspecies based on the biological species concept, though authors applying a genetic species concept have suggested it should be considered a species. Essentially restricted to the Himalayas, it differs little from the giant honey bee in appearance, but has extensive behavioral adaptations that enable it to nest in the open at high altitudes despite low ambient temperatures. It is the largest living honey bee.

Apis

The eastern species include three or four species. The reddish Koschevnikov's bee (*Apis koschevnikovi*) fromBorneo is well distinct; it probably derives from the first colonization of the island by cave-nesting honey bees. *Apis cerana*, the eastern honey bee proper, is the traditional honey bee of southern and eastern Asia, kept in hives in a similar fashion to *A. mellifera*, though on a much smaller and regionalised scale.

It has not been possible yet to resolve its relationship to the Bornean *A. c. nuluensis* and *Apis nigrocincta* from the Philippinesto satisfaction; the most recent hypothesis is that these are indeed distinct species, but that *A. cerana* is stillparaphyletic, consisting of several good species.

A. mellifera, the most common domesticated species, was the third insect to have its genome mapped. It seems to have originated in eastern tropical Africa and spread from there to Northern Europe and eastwards into Asia to the Tien Shan range. It is variously called the European, western or common honey bee in different parts of the world.

Many subspecies have adapted to the local geographic and climatic environments; in addition, hybrid strains, such as the Buckfast bee, have been bred. Behavior, color, and anatomy can be quite different from one subspecies or even strain to another.

Regarding phylogeny, this is the most enigmatic honey bee species. It seems to have diverged from its eastern relatives only during the Late Miocene. This would fit the hypothesis that the ancestral stock of cave-nesting honey bees was separated into the western group of East Africa and the eastern group of tropical Asia by desertification in the Middle East and adjacent regions, which caused declines of food plants and trees that provided nest sites, eventually causing gene flow to cease. The diversity of subspecies is probably the product of a largely Early Pleistocene radiation aided by climate and habitat changes during the last ice age.

That the western honey bee has been intensively managed by humans for many millennia – including hybridization and introductions – has apparently increased the speed of its evolution and confounded the DNA sequence data to a point where little of substance can be said about the exact relationships of many *A. mellifera* subspecies.

Apis mellifera is not native to the Americas, so was not present upon the arrival of the European explorers and colonists. However, other native bee species were kept and traded by indigenous peoples. In 1622, European colonists brought the dark bee (*A. m. mellifera*) to the Americas, followed later by Italian bees (*A. m. ligustica*) and others. Many of the crops that depend on honey bees for pollination have also been imported since colonial times.

Escaped swarms (known as "wild" bees, but actually feral) spread rapidly as far as the Great Plains, usually preceding the colonists. Honey bees did not naturally cross the Rocky Mountains; they were transported by the Mormon pioneers to Utah in the late 1840s, and by ship to California in the early 1850s.

Africanized bee

Africanized bees (known colloquially as "killer bees") are hybrids between European stock and one of the African subspecies *A. m. scutellata*; they are often more aggressive than European bees and do not create as much of a honey surplus, but are more resistant to disease and are better foragers. Originating by accident in Brazil, they have spread to North America and constitute a pest in some regions. However, these strains do not overwinter well, so are not often found in the colder, more northern parts of North America. The original breeding experiment for which the African bees were brought to Brazil in the first place has continued (though not as intended).

Novel hybrid strains of domestic and redomesticated Africanized bees combine high resilience to tropical conditions and good yields. They are popular among beekeepers in Brazil.

Beekeeping

Two species of honey bee, *A. mellifera* and *A. cerana indica*, are often maintained, fed, and transported by beekeepers. Modern hives also enable beekeepers to transport bees, moving from field to field as the crop needs pollinating and allowing the beekeeper to charge for the pollination services they provide, revising the historical role of the self-employed beekeeper, and favoring large-scale commercial operations.

Colony collapse disorder

Beekeepers in Western countries have been reporting slow declines of stocks for many years, apparently due to impaired protein production, changes in agricultural practice, or unpredictable weather. In early 2007, abnormally high die-offs (30–70% of hives) of European honey bee colonies occurred in North America; such a decline seems unprecedented in recent history.

This has been dubbed "colony collapse disorder" (CCD); it is unclear whether this is simply an accelerated phase of the general decline due to stochastically more adverse conditions in 2006, or a novel phenomenon. CCD is unique due to the lack of evidence as to what causes the sudden die off of adult worker bees, as well as there being few to no dead bees found around the hive.

Research has so far failed to determine what causes it, but the weight of evidence is tentatively leaning towards CCD being a syndromerather than a disease, as it seems to be caused by a combination of various contributing factors rather than a single pathogen or poison. However, in April 2013, after a report was released by the European Food Safety Authority identifying the significant risks of the class of pesticides called neonicotinoids, the European Union called for a two-year restriction on neonicotinoid pesticides.

A 2007 study linked CCD with Israel acute paralysis virus at a level of statistic significance. IAPV was found in 83. 3% of hives with CCD, and has a predictive value of 96. 1%, making it one of the most probable candidates as the infectious agent in CCD.

One other possible hypothesis is that the bees are falling victim to a combination of insecticides and parasites. Feral honey bees are prone to high levels of deformed wing virus (DWV).

The varroa mite thrives in honey bee colonies by sucking the hemolymph of honey bees, causing open wounds that are susceptible to varroosis. Higher levels of DWV are more prevalent in colonies that are not being treated for varroosis. Tobacco ring spot virus replicates (TRSV) spreads and negatively affects the health of honey bees indirectly. TRSV has a wide host range.

It can be transmitted from infected plant hosts, through parasites such as varroa mites, and ultimately affect insects like the honey bee. In January 2012, a researcher discovered *Apocephalus borealis* larvae, a parasitic fly known to prey on bumble bees and wasps, in a test tube containing a dead honey bee believed to have been affected by CCD.

There have been no effective preventative measures against CCD suggested to date.

Lifecycle

As in a few other types of eusocial bees, a colony generally contains one queen bee, a fertile female; seasonally up to a few thousand drone bees, or fertile males; and tens of thousands of sterile female worker bees. Details vary among the different species of honey bees, but common features include:

1. Eggs are laid singly in a cell in a wax honeycomb, produced and shaped by the worker bees. Using her spermatheca, the queen actually can choose to fertilize the egg she is laying, usually depending on into which cell she is laying. Drones develop from unfertilised eggs and are haploid, while females (queens and worker bees) develop from fertilised eggs and arediploid. Larvae are initially fed with royal jelly produced by worker bees, later switching to honey and pollen. The exception is a larva fed solely on royal jelly, which will develop into a queen bee. The larva undergoes several moultings before spinning a cocoon within the cell, and pupating.
2. Young worker bees, sometimes called "nurse bees", clean the hive and feed the larvae. When their royal jelly-producing glands begin to atrophy, they begin building comb cells. They progress to other within-colony tasks as they become older, such as receiving nectar and pollen from foragers, and guarding the hive. Later still, a worker takes her first orientation flights and finally leaves the hive and typically spends the remainder of her life as a forager.
3. Worker bees cooperate to find food and use a pattern of "dancing" (known as the bee dance or waggle dance) to communicate information regarding resources with each other; this dance varies from species to species, but all living species of *Apis* exhibit some form of the behavior. If the resources are very close to the hive, they may also exhibit a less specific dance commonly known as the "round dance".
4. Honey bees also perform tremble dances, which recruit receiver bees to collect nectar from returning foragers.
5. Virgin queens go on mating flights away from their home colony to a drone congregation area, and mate with multiple drones before returning. The drones die in the act of mating. Queen honey bees do not mate with drones from their home colony.
6. Colonies are established not by solitary queens, as in most bees, but by groups known as "swarms", which consist of a mated queen and a large contingent of worker bees. This group moves *en masse* to a nest site scouted by worker bees beforehand. Once they arrive, they immediately construct a new wax comb and begin to raise new worker brood. This type of nest founding is not seen in any other living bee

genus, though several groups of vespid wasps also found new nests by swarming (sometimes including multiple queens). Also, stingless beeswill start new nests with large numbers of worker bees, but the nest is constructed before a queen is escorted to the site, and this worker force is not a true "swarm".

WINTER SURVIVAL

In cold climates, honey bees stop flying when the temperature drops below about 10 °C (50 °F) and crowd into the central area of the hive to form a "winter cluster". The worker bees huddle around the queen bee at the center of the cluster, shivering to keep the center between 27 °C (81 °F) at the start of winter (during the broodless period) and 34 °C (93 °F) once the queen resumes laying. The worker bees rotate through the cluster from the outside to the inside so that no bee gets too cold.

The outside edges of the cluster stay at about 8–9 °C (46–48 °F). The colder the weather is outside, the more compact the cluster becomes. During winter, they consume their stored honey to produce body heat. The amount of honey consumed during the winter is a function of winter length and severity, but ranges in temperate climates from 15 to 50 kg (30 to 100 pounds).

Pollination

Species of *Apis* are generalist floral visitors, and will pollinate a large variety of plants, but by no means all plants. Of all the honey bee species, only *Apis mellifera* has been used extensively for commercial pollination of crops and other plants.

Fig. 3. Honey bee pollinating flowers

The value of these pollination services is commonly measured in the billions of dollars. Bees collect 66 pounds of pollen per year, per hive.

BEE PRODUCTS

Honey

Honey is the complex substance made when the nectar and sweet deposits from plants and trees are gathered, modified, and stored in the honeycomb by honey bees as a food source for the colony. All living species of *Apis* have had their honey gathered by indigenous peoples for consumption, though for commercial purposes, only *A. mellifera* and *A. cerana* have been utilized to any degree. Honey is sometimes also gathered by humans from the nests of various stingless bees.

In 1911, a bee culturist estimated a quart (about a litre) of honey represented bees flying over an estimated 48, 000 miles to gather the nectar needed to produce the honey.

Nectar

Nectar, a liquid high in sucrose, is produced in plant glands known as nectaries. It is an important energy resource for honey bees and plays a significant role in foraging economics and evolutionary differentiation between different subspecies. It was proposed through an experiment conducted with the African honey bee, *A. m. scutellata*, that nectar temperature impacts the foraging decisions of honey bees.

Beeswax

Worker bees of a certain age will secrete beeswax from a series of glands on their abdomens. They use the wax to form the walls and caps of the comb. As with honey, beeswax is gathered by humans for various purposes.

Pollen

Bees collect pollen in the pollen basket and carry it back to the hive. In the hive, pollen is used as a protein source necessary during brood-rearing. In certain environments, excess pollen can be collected from the hives of *A. mellifera* and *A. cerana*. It is often eaten as a health supplement. It also has been used with moderate success as a source of pollen for hand pollination However, pollen collected by bees and harvested for pollination must be used within a few hours because it loses its potency rapidly, possibly because of the effects of enzymes or other chemicals from the bees; hand-collected pollen may remain usable for weeks, if stored promptly under suitable conditions.

Propolis

Propolis or bee glue is created from resins, balsams, and tree saps. Those species of honey bees that nest in tree cavities use propolis to seal cracks in the hive. Dwarf honey bees use propolis to defend against ants by coating the branch from which their nest is suspended to create a sticky moat. Propolis is consumed by humans as a health supplement in various ways and also used in some cosmetics.

Sexes and castes

A caste is a different form, morphologically or reproductively, within the same sex of a species. Honey bees have three castes, drones, workers, and queens. There are two sexes: drones are male, while workers and queens are female.

Drones

Males or drones are typically haploid, having only one set of chromosomes. They are produced by the queen if she chooses not to fertilize an egg; or by an unfertilized laying worker. Diploid drones may be produced if an egg is fertilized but is homozygous for the sex-determination allele. Drones take 24 days to develop and may be produced from summer through autumn. Drones have large eyes used to locate queens during mating flights. Drones do not have a sting.

Workers

Workers are female bees and have two sets of chromosomes. They are produced from an egg that the queen has selectively fertilized from stored sperm. Workers typically develop in 21 days. A typical colony may contain as many as 60, 000 worker bees. Workers exhibit a wider range of behaviors than either queens or drones.

Their duties change upon the age of the bee in the following order (beginning with cleaning out their own cell after eating through their capped brood cell): feed brood, receive nectar, clean hive, guard duty, and foraging. Some workers engage in other specialized behaviors, such as "undertaking" (removing corpses of their nestmates from inside the hive).

Workers have morphological specializations, including the corbiculum or pollen basket, abdominal glands that produce beeswax, brood-feeding glands, and barbs on the sting. Under certain conditions (for example, if the colony becomes queenless), a worker may develop ovaries.

Queens

Queen honey bees, like workers, are female. They are created at the decision of the worker bees by feeding a larva only royal jelly throughout its development, rather than switching from royal jelly to pollen once the larva grows past a certain size. Queens are produced in oversized cells and develop in only 16 days.

Queens have a different morphology and behavior from worker bees. In addition to the greater size of the queen, she has a functional set of ovaries, and a spermatheca, which stores and maintains sperm after she has mated. The sting of queens is not barbed like a worker's sting, and queens lack the glands that produce beeswax. Once mated, queens may lay up to 2, 000 eggs per day. They produce a variety of pheromones that regulate behavior of workers, and helps swarms track the queen's location during the migratory phase.

Defense

All honey bees live in colonies where the workers sting intruders as a form of defense, and alarmed bees release a pheromone that stimulates the attack response in other bees. The different species of honey bees are distinguished from all other bee species (and virtually all other Hymenoptera) by the possession of small barbs on the sting, but these barbs are found only in the worker bees.

The sting and associated venom sac of honey bees are also modified so as to pull free of the body once lodged (autotomy), and the sting apparatus has its own musculature and ganglion, which allow it to keep delivering venom once detached. The worker dies after the sting becomes lodged and is subsequently torn loose from the bee's abdomen. The honey bee's venom, known as apitoxin, carries several active components, the most abundant of which is melittin, and the most destructive Phospholipase A2.

This complex apparatus, including the barbs on the sting, is thought to have evolved specifically in response to predation by vertebrates, as the barbs do not usually function (and the sting apparatus does not detach) unless the sting is embedded in fleshy tissue. While the sting can also penetrate the membranes between joints in the exoskeleton of other insects (and is used in fights between queens), in the case of *Apis cerana japonica*, defense against larger insects such as predatory wasps (e. g. Asian giant hornet) is usually performed by surrounding the intruder with a mass of defending worker bees, which vibrate their muscles vigorously to raise the temperature of the intruder to a lethal level. Previously, heat alone was thought to be responsible for killing intruding wasps, but recent experiments have demonstrated the increased temperature in combination with increased carbon dioxide levels within the ball produce the lethal effect. This phenomenon is also used to kill a queen perceived as intruding or defective, an action known to beekeepers as 'balling the queen', named for the ball of bees formed.

In the case of those honey bee species with open combs (e. g., *A. dorsata*), would-be predators are given a warning signal that takes the form of a "Mexican wave" that spreads as a ripple across a layer of bees densely packed on the surface of the comb when a threat is perceived, and consists of bees momentarily arching their bodies and flicking their wings.

Communication

Honey bees are known to communicate through many different chemicals and odors, as is common in insects, but also using specific behaviours that convey information about the quality and type of resources in the environment, and where these resources are located. The details of the signalling being used vary from species to species; for example, the two smallest species, *Apis andreniformis* and *A. florea*, dance on the upper surface of the comb, which is horizontal (not vertical, as in other species), and worker bees orient the dance in the actual compass direction of the resource to which they are recruiting.

Apis mellifera carnica honey bees use their antennae asymmetrically for social interactions with a strong lateral preference to use their right antenna.

Symbolism

Both the *Atharva Veda* and the ancient Greeks associated lips anointed with honey with the gift of eloquence and even of prescience. The priestess at Delphi was the "Delphic Bee". The Quran has a chapter titled "The Bee".

A community of honey bees has often been employed throughout history by political theorists as a model of human society:

This image occurs in Aristotle and Plato; in Virgil and Seneca; in Erasmus and Shakespeare; in Marx and Tolstoy.

Honey bees, signifying immortality and resurrection, were royal emblems of the Merovingians, revived by Napoleon. The bee also is the heraldic emblem of the Barberini.

8

Apiculture: A Beginners Guide

Beekeeping (or **apiculture**, from Latin: *apis* "bee") is the maintenance of honey bee colonies, commonly in hives, by humans. A beekeeper (or apiarist) keeps bees in order to collect theirhoney and other products that the hive produces (including beeswax, propolis, pollen, and royal jelly), to pollinate crops, or to produce bees for sale to other beekeepers. A location where bees are kept is called an apiary or "bee yard".

Depictions of humans collecting honey from wild bees date to 15, 000 years ago; efforts to domesticate them are shown in Egyptian art around 4, 500 years ago. Simple hives and smoke were used and honey was stored in jars, some of which were found in the tombs of pharaohssuch as Tutankhamun. It wasn't until the 18th century that European understanding of the colonies and biology of bees allowed the construction of the moveable comb hive so that honey could be harvested without destroying the entire colony.

HISTORY OF BEEKEEPING

At some point humans began to attempt to domesticate wild bees in artificial hives made from hollow logs, wooden boxes, pottery vessels, and woven straw baskets or "skeps". Honeybees were kept in Egypt from antiquity. On the walls of the sun temple of Nyuserre Ini from theFifth Dynasty, before 2422 BCE, workers are depicted blowing smoke into hives as they are removing honeycombs. Inscriptions detailing the production of honey are found on the tomb of Pabasa from the Twenty-sixth Dynasty (c. 650 BCE), depicting pouring honey in jars and cylindrical hives. Sealed pots of honey were found in the grave goods of pharaohs such as Tutankhamun.

There was an unsuccessful attempt to introduce bees to Mesopotamia in the 8th century BCE by Shamash-resh-u'ur, the governor of Mari and Suhu. His ambitious plans were detailed in a stele of 760 BCE:

I am Shamash-resh-u'ur, the governor of Suhu and the land of Mari. Bees that collect honey, which none of my ancestors had ever seen or brought into the land of Suhu, I brought down from the mountain of the men of Habha, and made them settle in the orchards of the town 'Gabbari-built-it'. They collect honey and wax, and I know how to melt the honey and wax - and the gardeners know too. Whoever comes in the future, may he ask the old men of the town, (who will say) thus: "They are the buildings of Shamash-resh-u'ur, the governor of Suhu, who introduced honey bees into the land of Suhu. "— translated text from stele, *(Dalley, 2002)*

In prehistoric Greece (Crete and Mycenae), there existed a system of high-status apiculture, as can be concluded from the finds of hives, smoking pots, honey extractors and other beekeeping paraphernalia in Knossos. Beekeeping was considered a highly valued industry controlled by beekeeping overseers—owners of gold rings depicting apiculture scenes rather than religious ones as they have been reinterpreted recently, contra Sir Arthur Evans.

Archaeological finds relating to beekeeping have been discovered at Rehov, a Bronze and Iron Age archaeological site in the Jordan Valley, Israel. Thirty intact hives, made of straw and unbaked clay, were discovered by archaeologist Amihai Mazar of the Hebrew University of Jerusalem in the ruins of the city, dating from about 900 BCE.

The hives were found in orderly rows, three high, in a manner that could have accommodated around 100 hives, held more than 1 million bees and had a potential annual yield of 500 kilograms of honey and 70 kilograms of beeswax, according to Mazar, and are evidence that an advanced honey industry existed in ancient Israel 3, 000 years ago. Ezra Marcus, an expert from the University of Haifa, said the finding was a glimpse of ancient beekeeping seen in texts and ancient art from the Near East.

In ancient Greece, aspects of the lives of bees and beekeeping are discussed at length by Aristotle. Beekeeping was also documented by the Roman writers Virgil, Gaius Julius Hyginus, Varro, and Columella.

Beekeeping has also been practiced in ancient China since antiquity. In the book "Golden Rules of Business Success" written by Fan Li (or Tao Zhu Gong) during the Spring and Autumn Period there are sections describing the art of beekeeping, stressing the importance of the quality of the wooden box used and how this can affect the quality of the honey.

Origins

There are more than 20, 000 species of wild bees. Many species are solitary (e. g., mason bees, leafcutter bees (Megachilidae), carpenter bees and other ground-nesting bees). Many others rear their young in burrows and small colonies (e. g., bumblebees and stingless bees). Some honey bees are wild e. g. the little honeybee (*Apis florea*), giant honeybee (*Apis dorsata*) and rock bee (*Apis laboriosa*).

Beekeeping, or apiculture, is concerned with the practical management of the social species of honey bees, which live in large colonies of up to 100, 000 individuals. In Europe and America the species universally managed by beekeepers is the Western honey bee (*Apis mellifera*). This species has several sub-species or regional varieties, such as the Italian bee (*Apis mellifera ligustica*), European dark bee (*Apis mellifera mellifera*), and the Carniolan honey bee (*Apis mellifera carnica*).

In the tropics, other species of social bees are managed for honey production, including the Asiatic honey bee (*Apis cerana*).

All of the *Apis mellifera* sub-species are capable of inter-breeding and hybridizing. Many bee breeding companies strive to selectively breed and hybridize varieties to produce desirable qualities: disease and parasite resistance, good honey production, swarming behaviour reduction, prolific breeding, and mild disposition. Some of these hybrids are marketed under specific brand names, such as the Buckfast Bee or Midnite Bee.

The advantages of the initial F1 hybrids produced by these crosses include: hybrid vigor, increased honey productivity, and greater disease resistance. The disadvantage is that in subsequent generations these advantages may fade away and hybrids tend to be very defensive and aggressive.

Wild honey harvesting

Collecting honey from wild bee colonies is one of the most ancient human activities and is still practiced by aboriginal societies in parts of Africa, Asia, Australia, and South America. Some of the earliest evidence of gathering honey from wild colonies is from rock paintings, dating to around 13, 000 BCE. Gathering honey from wild bee colonies is usually done by subduing the bees with smoke and breaking open the tree or rocks where the colony is located, often resulting in the physical destruction of the nest location.

Fig. 1. Natural beehive

Study of honey bees

It was not until the 18th century that European natural philosophers undertook the scientific study of bee colonies and began to understand the complex and hidden world of bee biology. Preeminent among these scientific pioneers wereSwammerdam, René Antoine Ferchault de Réaumur, Charles Bonnet, and Francois Huber.

Swammerdam and Réaumur were among the first to use a microscope and dissection to understand the internal biology of honey bees. Réaumur was among the first to construct a glass walled observation hive to better observe activities within hives.

He observed queens laying eggs in open cells, but still had no idea of how a queen was fertilized; nobody had ever witnessed the mating of a queen and drone and many theories held that queens were "self-fertile, " while others believed that a vapor or "miasma" emanating from the drones fertilized queens without direct physical contact.

Huber was the first to prove by observation and experiment that queens are physically inseminated by drones outside the confines of hives, usually a great distance away.

Following Réaumur's design, Huber built improved glass-walled observation hives and sectional hives that could be opened like the leaves of a book. This allowed inspecting individual wax combs and greatly improved direct observation of hive activity.

Although he went blind before he was twenty, Huber employed a secretary, Francois Burnens, to make daily observations, conduct careful experiments, and keep accurate notes over more than twenty years. Huber confirmed that a hive consists of one queen who is the mother of all the female workers and male drones in the colony.

He was also the first to confirm that mating with drones takes place outside of hives and that queens are inseminated by a number of successive matings with male drones, high in the air at a great distance from their hive. Together, he and Burnens dissected bees under the microscope and were among the first to describe the ovaries and spermatheca, or sperm store, of queens as well as the penisof male drones.

Huber is universally regarded as "the father of modern bee-science" and his "Nouvelles Observations sur Les Abeilles (or "New Observations on Bees") revealed all the basic scientific truths for the biology and ecology of honeybees.

Invention of the movable comb hive

Early forms of honey collecting entailed the destruction of the entire colony when the honey was harvested. The wild hive was crudely broken into, using smoke to suppress the bees, the honeycombswere torn out and smashed up — along with the eggs, larvae and honey they contained. The liquid honey from the destroyed brood nest was strained through a sieve or basket.

This was destructive and unhygienic, but for hunter-gatherer societies this did not matter, since the honey was generally consumed immediately and there were always more wild colonies to exploit. But in settled societies the destruction of the bee colony meant the loss of a valuable resource; this drawback made beekeeping both inefficient and something of a "stop and start" activity.

There could be no continuity of production and no possibility of selective breeding, since each bee colony was destroyed at harvest time, along with its precious queen.

During the medieval period abbeys and monasteries were centers of beekeeping, since beeswax was highly prized for candles and fermented honey was used to make alcoholic mead in areas of Europe where vines would not grow. The 18th and 19th centuries saw successive stages of a revolution in beekeeping, which allowed the bees themselves to be preserved when taking the harvest.

Intermediate stages in the transition from the old beekeeping to the new were recorded for example by Thomas Wildman in 1768/1770, who described advances over the destructive old skep-based beekeeping so that the bees no longer had to be killed to harvest the honey.

Wildman for example fixed a parallel array of wooden bars across the top of a straw hive or skep (with a separate straw top to be fixed on later) "so that there are in all seven bars of deal" [in a 10-inch-diameter (250 mm) hive] "to which the bees fix their combs. "

He also described using such hives in a multi-storey configuration, foreshadowing the modern use of supers: he described adding (at a proper time) successive straw hives below, and eventually removing the ones above when free of brood and filled with honey, so that the bees could be separately preserved at the harvest for a following season. Wildman also described a further development, using hives with "sliding frames" for the bees to build their comb, foreshadowing more modern uses of movable-comb hives.

Wildman's book acknowledged the advances in knowledge of bees previously made by Swammerdam, Maraldi, and de Réaumur—he included a lengthy translation of Réaumur's account of the natural history of bees—and he also described the initiatives of others in designing hives for the preservation of bee-life when taking the harvest, citing in particular reports from Brittany dating from the 1750s, due to Comte de la Bourdonnaye.

However, the forerunners of the modern hives with movable frames that are mainly used today are considered the traditional basket top bar (movable comb) hives of Greece, known as "Greek beehives". The oldest testimony on their use dates back to 1669 although it is probable that their use is more than 3000 years old. The 19th century saw this revolution in beekeeping practice completed through the perfection of the movable comb hive by the American Lorenzo Lorraine Langstroth.

Langstroth was the first person to make practical use of Huber's earlier discovery that there was a specific spatial measurement between the wax combs, later called *the bee space*, which bees do not block with wax, but keep as a free passage.

Having determined this bee space (between 5 and 8 mm, or 1/4 to 3/8"), Langstroth then designed a series of wooden frames within a rectangular hive box, carefully maintaining the correct space between successive frames, and found that the bees would build parallel honeycombs in the box without bonding them to each other or to the hive walls. This enables the beekeeper to slide any frame out of the hive for inspection, without harming the bees or the comb, protecting the eggs, larvae and pupae contained within the cells. It also meant that combs containing honey could be gently removed and the honey extracted without destroying the comb. The emptied honey combs could then be returned to the bees intact for refilling. Langstroth's classic book, *The Hive and Honey-bee*, published in 1853, described his rediscovery of the bee space and the development of his patent movable comb hive.

The invention and development of the movable-comb-hive fostered the growth of commercial honey production on a large scale in both Europe and the USA.

Evolution of hive designs

Langstroth's design for movable comb hives was seized upon by apiarists and inventors on both sides of the Atlantic and a wide range of moveable comb hives were designed and perfected in England, France, Germany and the United States. Classic designs evolved in each country: Dadant hives and Langstroth hives are still dominant in the USA; in France the De-Layens trough-hive became popular and in the UK a British National hive became standard as late as the 1930s although in Scotland the smaller Smith hive is still popular.

Fig. 2. Honey-laden honeycomb in a wooden frame

In some Scandinavian countries and in Russia the traditional trough hive persisted until late in the 20th century and is still kept in some areas.

However, the Langstroth and Dadant designs remain ubiquitous in the USA and also in many parts of Europe, though Sweden, Denmark, Germany, France and Italy all have their own national hive designs. Regional variations of hive evolved to reflect the climate, floral productivity and the reproductive characteristics of the various subspecies of native honey bee in each bio-region.

The differences in hive dimensions are insignificant in comparison to the common factors in all these hives: they are all square or rectangular; they all use movable wooden frames; they all consist of a floor, brood-box, honey super, crown-board and roof. Hives have traditionally been constructed ofcedar, pine, or cypress wood, but in recent years hives made from injection molded dense polystyrenehave become increasingly important.

Hives also use queen excluders between the brood-box and honey supers to keep the queen from laying eggs in cells next to those containing honey intended for consumption. Also, with the advent in the 20th century of mite pests, hive floors are often replaced for part of (or the whole) year with a wire mesh and removable tray.

Pioneers of practical and commercial beekeeping

The 19th century produced an explosion of innovators and inventors who perfected the design and production of beehives, systems of management and husbandry, stock improvement by selective breeding, honey extraction and marketing. Preeminent among these innovators were:

Petro Prokopovych, used frames with channels in the side of the woodwork, these were packed side by side in boxes that were stacked one on top of the other. The bees travelling from frame to frame and box to box via the channels. The channels were similar to the cut outs in the sides of modern wooden sections (1814).

Jan Dzier¿on, was the father of modern apiology and apiculture. All modern beehives are descendants of his design.

L. L. Langstroth, revered as the "father of American apiculture", no other individual has influenced modern beekeeping practice more than Lorenzo Lorraine Langstroth. His classic book *The Hive and Honey-bee* was published in 1853.

Moses Quinby, often termed 'the father of commercial beekeeping in the United States', author of *Mysteries of Bee-Keeping Explained*.

Amos Root, author of the *A B C of Bee Culture*, which has been continuously revised and remains in print. Root pioneered the manufacture of hives and the distribution of bee-packages in the United States.

A. J. Cook, author of *The Bee-Keepers' Guide; or Manual of the Apiary*, 1876.

Dr. C. C. Miller was one of the first entrepreneurs to actually make a living from apiculture. By 1878 he made beekeeping his sole business activity. His book, *Fifty Years Among the Bees*, remains a classic and his influence on bee management persists to this day.

Major Francesco De Hruschka was an Italian military officer who made one crucial invention that catalyzed the commercial honey industry. In 1865 he invented a simple machine for extracting honey from the comb by means of centrifugal force. His original idea was simply to support combs in a metal framework and then spin them around within a container to collect honey as it was thrown out by centrifugal force. This meant that honeycombs could be returned to a hive undamaged but empty, saving the bees a vast amount of work, time, and materials. This single invention greatly improved the efficiency of honey harvesting and catalysed the modern honey industry.

Walter T. Kelley was an American pioneer of modern beekeeping in the early and mid-20th century. He greatly improved upon beekeeping equipment and clothing and went on to manufacture these items as well as other equipment. His company sold via catalog worldwide and his book, *How to Keep Bees & Sell Honey*, an introductory book of apiculture and marketing, allowed for a boom in beekeeping following World War II.

In the U. K. practical beekeeping was led in the early 20th century by a few men, pre-eminently **Brother Adam** and his Buckfast bee and**R. O. B. Manley**, author of many titles, including *Honey Production in the British Isles* and inventor of the Manley frame, still universally popular in the U. K. Other notable British pioneers include William Herrod-Hempsall and Gale.

Dr. Ahmed Zaky Abushady (1892–1955), was an Egyptian poet, medical doctor, bacteriologist and bee scientist who was active in England and in Egypt in the early part of the twentieth century. In 1919, Abushady patented a removable, standardized aluminum honeycomb. In 1919 he also founded The Apis Club in Benson, Oxfordshire, and its periodical

Bee World, which was to be edited by Annie D. Betts and later by Dr. Eva Crane. The Apis Club was transitioned to the International Bee Research Association (IBRA). Its archives are held in the National Library of Wales. In Egypt in the 1930s, Abushady established The Bee Kingdom League and its organ, The Bee Kingdom.

In India, R. N. Mattoo was the pioneer worker in starting beekeeping with Indian honeybee, (*Apis cerana indica*) in early 1930s. Beekeeping with European honeybee, (*Apis mellifera*) was started by **Dr. A. S. Atwal** and his team members, **O. P. Sharma** and **N. P. Goyal** in Punjab in early 1960s. It remained confined to Punjab and Himachal Pradesh up to late 1970s. Later on in 1982, **Dr. R. C. Sihag**, working at Haryana Agricultural University, Hisar (Haryana), introduced and established this honeybee in Haryana and standardized its management practices for semi-arid-subtropical climates. On the basis of these practices, beekeeping with this honeybee could be extended to the rest of the country. Now beekeeping with *Apis mellifera* predominates in India.

Traditional beekeeping

Fixed comb hives

A fixed comb hive is a hive in which the combs cannot be removed or manipulated for management or harvesting without permanently damaging the comb. Almost any hollow structure can be used for this purpose, such as a log gum, skep or a clay pot. Fixed comb hives are no longer in common use in industrialized countries, and are illegal in some places that require inspection for problems such asvarroa and American foulbrood. In many developing countries fixed comb hives are widely used and, because they can be made from any locally available material, are very inexpensive. Beekeeping using fixed comb hives is an essential part of the livelihoods of many communities in poor countries. The charity Bees for Development recognizes that local skills to manage bees in fixed comb hives are widespread in Africa, Asia, and South America.

Modern beekeeping

Movable frame hives

In the United States, the Langstroth hive is commonly used. The Langstroth was the first successful top-opened hive with movable frames, and other designs of hive have been based on it. The Langstroth hive was, however, a descendant of Jan Dzierzon's Polish hive designs.

In the United Kingdom, the most common type of hive is the British National Hive, which can hold Hoffman, British Standard or popular Manley frames, but it is not unusual to see some other sorts of hive (Smith, Commercial and WBC, rarely Langstroth). Straw skeps, bee gums, and unframed box hives are now unlawful in most US states, as the comb and brood cannot be inspected for diseases. However, straw skeps are still used for collecting swarms by hobbyists in the UK, before moving them into standard hives.

Top-bar hives

A growing number of amateur beekeepers are adopting various top-bar hives similar to the type commonly found in Africa. Top bar hives were originally used as a traditional beekeeping method in Greece and Vietnam. These hives have no frames and the honey-filled comb is not returned after extraction, as it is in the Langstroth hive.

Because of this, the production of honey is likely to be somewhat less than that of a Langstroth hive. Top bar hives are mostly kept by people who are more interested in having bees in their garden than in honey production per se.

Some of the most well known top-bar hives are the Kenyan Top Bar Hive with sloping sides, the Tanzanian Top Bar Hive with straight sides, and Vertical Top Bar Hives, such as the Warre or "People's Hive" designed by Abbe Warre in the mid-1900s.

The initial costs and equipment requirements are far less. Scrap wood or #2 or #3 pine can often be used build a nice hive. Top-bar hives also offer some advantages to interacting with the bees and the amount of weight that must be lifted is greatly reduced.

Top-bar hives are being widely used in developing countries in Africa and Asia as a result of the Bees for Development program. Since 2011, a growing number of beekeepers in the U. S. are using various top-bar hives.

Protective clothing

Most beekeepers also wear some protective clothing. Novice beekeepers usually wear gloves and a hooded suit or hat and veil. Experienced beekeepers sometimes elect not to use gloves because they inhibit delicate manipulations. The face and neck are the most important areas to protect, so most beekeepers wear at least a veil.

Defensive bees are attracted to the breath, and a sting on the face can lead to much more pain and swelling than a sting elsewhere, while a sting on a bare hand can usually be quickly removed by fingernail scrape to reduce the amount of venom injected.

The protective clothing is generally light colored (but not colorful) and of a smooth material. This provides the maximum differentiation from the colony's natural predators.

'Stings' retained in clothing fabric continue to pump out an alarm pheromone that attracts aggressive action and further stinging attacks. Washing suits regularly, and rinsing gloved hands in vinegar minimizes attraction.

Smoker

Smoke is the beekeeper's third line of defense. Most beekeepers use a "smoker"—a device designed to generate smoke from the incomplete combustion of various fuels. Smoke calms bees; it initiates a feeding response in anticipation of possible hive abandonment due to fire. Smoke also masks alarm pheromones released by guard bees or when bees are squashed in an inspection.

The ensuing confusion creates an opportunity for the beekeeper to open the hive and work without triggering a defensive reaction. In addition, when a bee consumes honey the bee's abdomen distends, supposedly making it difficult to make the necessary flexes to sting, though this has not been tested scientifically.

Smoke is of questionable use with a swarm, because swarms do not have honey stores to feed on in response. Usually smoke is not needed, since swarms tend to be less defensive, as they have no stores to defend, and a fresh swarm has fed well from the hive. Many types of fuel can be used in a smoker as long as it is natural and not contaminated with harmful substances. These fuels include hessian, twine, burlap, pine needles, corrugated cardboard, and mostly rotten or punky wood.

Indian beekeepers, especially in Kerala, often use coconut fibers as they are readily available, safe, and of negligible expense. Some beekeeping supply sources also sell commercial fuels like pulped paper and compressed cotton, or even aerosol cans of smoke. Other beekeepers usesumac as fuel because it ejects lots of smoke and doesn't have an odor.

Some beekeepers are using "liquid smoke" as a safer, more convenient alternative. It is a water-based solution that is sprayed onto the bees from a plastic spray bottle.

Torpor may also be induced by the introduction of chilled air into the hive - while chilled carbon dioxide may have harmful long-term effects.

Effects of stings and of protective measures

Some beekeepers believe that the more stings a beekeeper receives, the less irritation each causes, and they consider it important for safety of the beekeeper to be stung a few times a season. Beekeepers have high levels of antibodies (mainly IgG) reacting to the major antigen ofbee venom, phospholipase A2 (PLA). Antibodies correlate with the frequency of bee stings.

The entry of venom into the body from bee-stings may also be hindered and reduced by protective clothing that allows the wearer to remove stings and venom sacs with a simple tug on the clothing. Although the stinger is barbed, a worker bee is less likely to become lodged into clothing than human skin.

Natural beekeeping

The natural beekeeping movement believes that modern beekeeping and agricultural practices, such as crop spraying, hive movement, frequent hive inspections, artificial insemination of queens, routine medication, and sugar water feeding, weaken bee hives.

Practitioners of 'natural beekeeping' tend to use variations of the top-bar hive, which is a simple design that retains the concept of movable comb without the use of frames or foundation. The horizontal top-bar hive, as championed by Marty Hardison, Michael Bush, Philip Chandler, Dennis Murrell and others, can be seen as a modernization of hollow log hives, with the addition of wooden bars of specific width from which bees hang their combs. Its widespread adoption in recent years can be attributed to the publication in 2007 of *The Barefoot Beekeeper* by Philip Chandler, which challenged many aspects of modern beekeeping and offered the horizontal top-bar hive as a viable alternative to the ubiquitous Langstroth-style movable-frame hive.

The most popular vertical top-bar hive is probably the Warré hive, based on a design by the French priest Abbé Émile Warré (1867–1951) and popularized by Dr. David Heaf in his English translation of Warré's book *L'Apiculture pour Tous* as *Beekeeping For All*.

Urban or backyard beekeeping

Related to natural beekeeping, urban beekeeping is an attempt to revert to a less industrialized way of obtaining honey by utilizing small-scale colonies that pollinate urban gardens. Urban apiculture has undergone a renaissance in the first decade of the 21st century, and urban beekeeping is seen by many as a growing trend; it has recently been legalized in cities where it was previously banned. Paris, Berlin, London, Tokyo, Melbourne, New York, and Washington DC are among beekeeping cities.

Some have found that "city bees" are actually healthier than "rural bees" because there are fewer pesticides and greater biodiversity. Urban bees may fail to find forage, however, and homeowners can use their landscapes to help feed local bee populations by planting flowers that provide nectar and pollen. An environment of year-round, uninterrupted bloom creates an ideal environment for colony reproduction.

Bee colonies

Castes

A colony of bees consists of three castes of bee:

- a queen bee, which is normally the only breeding female in the colony;
- a large number of female worker bees, typically 30, 000–50, 000 in number;
- a number of male drones, ranging from thousands in a strong hive in spring to very few during dearth or cold season.

Fig. 3. Queen bee (center)

The queen is the only sexually mature female in the hive and all of the female worker bees and male drones are her offspring. The queen may live for up to three years or more and may be capable of laying half a million eggs or more in her lifetime.

At the peak of the breeding season, late spring to summer, a good queen may be capable of laying 3, 000 eggs in one day, more than her own body weight. This would be exceptional however; a prolific queen might peak at 2, 000 eggs a day, but a more average queen might lay just 1, 500 eggs per day.

The queen is raised from a normal worker egg, but is fed a larger amount of royal jelly than a normal worker bee, resulting in a radically different growth and metamorphosis. The queen influences the colony by the production and dissemination of a variety of pheromones or "queen substances". One of these chemicals suppresses the development of ovaries in all the female worker bees in the hive and prevents them from laying eggs.

Mating of queens

The queen emerges from her cell after 15 days of development and she remains in the hive for 3–7 days before venturing out on a mating flight. Mating flight is otherwise known as 'nuptial flight'. Her first orientation flight may only last a few seconds, just enough to mark the position of the hive.

Subsequent mating flights may last from 5 minutes to 30 minutes, and she may mate with a number of male drones on each flight. Over several matings, possibly a dozen or more, the queen receives and stores enough sperm from a succession of drones to fertilize hundreds of thousands of eggs.

If she does not manage to leave the hive to mate—possibly due to bad weather or being trapped in part of the hive—she remains infertile and become a *drone layer*, incapable of producing female worker bees. Worker bees sometimes kill a non-performing queen and produce another. Without a properly performing queen, the hive is doomed.

Mating takes place at some distance from the hive and often several hundred feet in the air; it is thought that this separates the strongest drones from the weaker ones, ensuring that only the fastest and strongest drones get to pass on their genes.

Worker bees

Almost all the bees in a hive are female worker bees. At the height of summer when activity in the hive is frantic and work goes on non-stop, the life of a worker bee may be as short as 6 weeks; in late autumn, when no brood is being raised and no nectar is being harvested, a young bee may live for 16 weeks, right through the winter.

Fig. 4. Female worker bee

During its life a worker bee performs different work functions in the hive, largely dictated by the age of the bee.

Period	Work activity
Days 1-3	Cleaning cells and incubation
Day 3-6	Feeding older larvae
Day 6-10	Feeding younger larvae
Day 8-16	Receiving nectar and pollen from field bees
Day 12-18	Beeswax making and cell building
Day 14 onwards	Entrance guards; nectar, pollen, water and propolis foraging; robbing other hives

Drones

Drones are the largest bees in the shive (except for the queen), at almost twice the size of a worker bee. They do not work, do not forage for pollen or nectar and have no other known function than to mate with new queens and fertilize them on their mating flights.

Fig. 5. Larger drones compared to smaller workers

A bee colony generally starts to raise drones a few weeks before building queen cells so they can supersede a failing queen or prepare for swarming. When queen-raising for the season is over, bees in colder climates drive drones out of the hive to die, biting and tearing their legs and wings.

STRUCTURE OF A BEE COLONY

A domesticated bee colony is normally housed in a rectangular hive body, within which eight to ten parallel frames house the vertical plates of honeycomb that contain the eggs, larvae, pupae and food for the colony. If one were to cut a vertical cross-section through the hive from side to side, the brood nest would appear as a roughly ovoid ball spanning 5-8 frames of comb. The two outside combs at each side of the hive tend to be exclusively used for long-term storage of honey and pollen.

Within the central brood nest, a single frame of comb typically has a central disk of eggs, larvae and sealed brood cells that may extend almost to the edges of the frame.

Immediately above the brood patch an arch of pollen-filled cells extends from side to side, and above that again a broader arch of honey-filled cells extends to the frame tops. The pollen is protein-rich food for developing larvae, while honey is also food but largely energy rich rather than protein rich. The nurse bees that care for the developing brood secrete a special food called 'royal jelly' after feeding themselves on honey and pollen. The amount of royal jelly fed to a larva determines whether it develops into a worker bee or a queen.

Apart from the honey stored within the central brood frames, the bees store surplus honey in combs above the brood nest. In modern hives the beekeeper places separate boxes, called 'supers', above the brood box, in which a series of shallower combs is provided for storage of honey. This enables the beekeeper to remove some of the supers in the late summer, and to extract the surplus honey harvest, without damaging the colony of bees and its brood nest below. If all the honey is 'stolen', including the amount of honey needed to survive winter, the beekeeper must replace these stores by feeding the bees sugar or corn syrup in autumn.

Annual cycle of a bee colony

The development of a bee colony follows an annual cycle of growth that begins in spring with a rapid expansion of the brood nest, as soon as pollen is available for feeding larvae. Some production of brood may begin as early as January, even in a cold winter, but breeding accelerates towards a peak in May (in the northern hemisphere), producing an abundance of harvesting bees synchronized to the main nectar flow in that region. Each race of bees times this build-up slightly differently, depending on how the flora of its original region blooms. Some regions of Europe have two nectar flows: one in late spring and another in late August. Other regions have only a single nectar flow. The skill of the beekeeper lies in predicting when the nectar flow will occur in his area and in trying to ensure that his colonies achieve a maximum population of harvesters at exactly the right time.

The key factor in this is the prevention or skillful management of the swarming impulse. If a colony swarms unexpectedly and the beekeeper does not manage to capture the resulting swarm, he is likely to harvest significantly less honey from that hive, since he has lost half his worker bees at a single stroke. If, however, he can use the swarming impulse to breed a new queen but keep all the bees in the colony together, he maximizes his chances of a good harvest. It takes many years of learning and experience to be able to manage all these aspects successfully, though owing to variable circumstances many beginners often achieve a good honey harvest.

Formation of new colonies

Colony reproduction: swarming and supersedure

All colonies are totally dependent on their queen, who is the only egg-layer. However, even the best queens live only a few years and one or two years longevity is the norm. She can choose whether or not to fertilize an egg as she lays it; if she does so, it develops into a female worker bee; if she lays an unfertilized egg it becomes a male drone. She decides which type of egg to lay depending on the size of the open brood cell she encounters on the comb. In a small worker cell, she lays a fertilized egg; if she finds a larger drone cell, she lays an unfertilized drone egg.

All the time that the queen is fertile and laying eggs she produces a variety of pheromones, which control the behavior of the bees in the hive. These are commonly called *queen substance*, but there are various pheromones with different functions. As the queen ages, she begins to run out of stored sperm, and her pheromones begin to fail. Inevitably, the queen begins to falter, and the bees decide to replace her by creating a new queen from one of her worker eggs. They may do this because she has been damaged (lost a leg or an antenna), because she has run out of sperm and cannot lay fertilized eggs (has become a 'drone laying queen'), or because her pheromones have dwindled to where they cannot control all the bees in the hive.

At this juncture, the bees produce one or more queen cells by modifying existing worker cells that contain a normal female egg. However, the bees pursue two distinct behaviors:

1. Supersedure: queen replacement within one hive without swarming
2. Swarm cell production: the division of the hive into two colonies by swarming

Different sub-species of *Apis mellifera* exhibit differing swarming characteristics that reflect their evolution in different ecotopes of the European continent. In general the more northerly black races are said to swarm less and supersede more, whereas the more southerly yellow and grey varieties are said to swarm more frequently. The truth is complicated because of the prevalence of cross-breeding and hybridization of the sub species and opinions differ.

Supersedure is highly valued as a behavioral trait by beekeepers because a hive that supersedes its old queen does not swarm and so no stock is lost; it merely creates a new queen and allows the old one to fade away, or alternatively she is killed when the new queen emerges. When superseding a queen, the bees produce just one or two queen cells, characteristically in the center of the face of a broodcomb.

In swarming, by contrast, a great many queen cells are created—typically a dozen or more—and these are located around the edges of a broodcomb, most often at the sides and the bottom.

Once either process has begun, the old queen normally leaves the hive with the hatching of the first queen cells. She leaves accompanied by a large number of bees, predominantly young bees (wax-secretors), who form the basis of the new hive. Scouts are sent out from the swarm to find suitable hollow trees or rock crevices. As soon as one is found, the entire swarm moves in. Within a matter of hours, they build new wax brood combs, using honey stores that the young bees have filled themselves with before leaving the old hive. Only young bees can secrete wax from special abdominal segments, and this is why swarms tend to contain more young bees. Often a number of virgin queens accompany the first swarm (the 'prime swarm'), and the old queen is replaced as soon as a daughter queen mates and begins laying. Otherwise, she is quickly superseded in the new home.

Factors that trigger swarming

It is generally accepted that a colony of bees does not swarm until they have completed all of their brood combs, i. e., filled all available space with eggs, larvae, and brood. This generally occurs in late spring at a time when the other areas of the hive are rapidly filling with honey stores. One key trigger of the swarming instinct is when the queen has no more room to lay eggs and the hive population is becoming very congested. Under these conditions, a prime swarm may issue with the queen, resulting in a halving of the population within the hive, leaving the old colony with a large number of hatching bees. The queen who leaves finds herself in a new hive with no eggs and no larvae but lots of energetic young bees who create a new set of brood combs from scratch in a very short time.

Another important factor in swarming is the age of the queen. Those under a year in age are unlikely to swarm unless they are extremely crowded, while older queens have swarming predisposition.

Beekeepers monitor their colonies carefully in spring and watch for the appearance of queen cells, which are a dramatic signal that the colony is determined to swarm.

When a colony has decided to swarm, queen cells are produced in numbers varying to a dozen or more. When the first of these queen cells is sealed after eight days of larval feeding, a virgin queen pupates and is due to emerge seven days later. Before leaving, the worker bees fill their stomachs with honey in preparation for the creation of new honeycombs in a new home. This cargo of honey also makes swarming bees less inclined to sting. A newly issued swarm is noticeably gentle for up to 24 hours and is often capable of being handled by a beekeeper without gloves or veil. This swarm looks for shelter. A beekeeper may capture it and introduce it into a new hive, helping meet this need. Otherwise, it returns to a feral state, in which case it finds shelter in a hollow tree, excavation, abandoned chimney, or even behind shutters.

Back at the original hive, the first virgin queen to emerge from her cell immediately seeks to kill all her rival queens still waiting to emerge. Usually, however, the bees deliberately prevent her from doing this, in which case, she too leads a second swarm from the hive. Successive swarms are called 'after-swarms' or 'casts' and can be very small, often with just a thousand or so bees—as opposed to a prime swarm, which may contain as many as ten to twenty-thousand bees.

A small after-swarm has less chance of survival and may threaten the original hive's survival if the number of individuals left is unsustainable. When a hive swarms despite the beekeeper's preventative efforts, a good management practice is to give the reduced hive a couple frames of open brood with eggs. This helps replenish the hive more quickly and gives a second opportunity to raise a queen if there is a mating failure.

Each race or sub-species of honey bee has its own swarming characteristics. Italian bees are very prolific and inclined to swarm; Northern European black bees have a strong tendency to supersede their old queen without swarming. These differences are the result of differing evolutionary pressures in the regions where each sub-species evolved.

Artificial swarming

When a colony accidentally loses its queen, it is said to be "queenless". The workers realize that the queen is absent after as little as an hour, as her pheromones fade in the hive. The colony cannot survive without a fertile queen laying eggs to renew the population, so the workers select cells containing eggs aged less than three days and enlarge these cells dramatically to form "emergency queen cells". These appear similar to large peanut-like structures about an inch long that hang from the center or side of the brood combs. The developing larva in a queen cell is fed differently from an ordinary worker-bee; in addition to the normal honey and pollen, she receives a great deal of royal jelly, a special food secreted by young 'nurse bees' from the hypopharyngeal gland. This special food dramatically alters the growth and development of the larva so that, after metamorphosis and pupation, it emerges from the cell as a queen bee. The queen is the only bee in a colony which has fully developed ovaries, and she secretes a pheromone which suppresses the normal development of ovaries in all her workers.

Beekeepers use the ability of the bees to produce new queens to increase their colonies in a procedure called *splitting a colony*. To do this, they remove several brood combs from a healthy hive, taking care to leave the old queen behind. These combs must contain eggs or larvae less than three days old and be covered by young *nurse bees*, which care for the brood and keep it warm. These brood combs and attendant nurse bees are then placed into a small 'nucleus hive' with other combs containing honey and pollen. As soon as the nurse bees find themselves in this new hive and realize they have no queen, they set about constructing emergency queen cells using the eggs or larvae they have in the combs with them.

Diseases

The common agents of disease that affect adult honey bees include fungi, bacteria, protozoa, viruses, parasites, and poisons. The gross symptoms displayed by affected adult bees are very similar, whatever the cause, making it difficult for the apiarist to ascertain the causes of problems without microscopic identification of microorganisms or chemical analysis of poisons. Since 2006 colony losses from Colony Collapse Disorder have been increasing across the world although the causes of the syndrome are, as yet, unknown.

9

Guide to Keeping Bees and Making Honey

Bees are flying insects closely related to wasps and ants, and are known for their role in pollinationand for producing honey and beeswax. Bees are a monophyletic lineage within the superfamilyApoidea, presently considered as a clade **Anthophila**. There are nearly 20, 000 known species of bees in seven to nine recognized families, though many are undescribed and the actual number is probably higher. They are found on every continent except Antarctica, in every habitat on the planet that contains insect-pollinated flowering plants.

Bees are adapted for feeding on nectar and pollen, the former primarily as an energy source and the latter primarily for protein and other nutrients. Most pollen is used as food for larvae.

Bees have a long proboscis (a complex "tongue") that enables them to obtain the nectar from flowers. They have antennae almost universally made up of 13 segments in males and 12 in females, as is typical for the superfamily. Bees all have two pairs of wings, the hind pair being the smaller of the two; in a very few species, one sex or caste has relatively short wings that make flight difficult or impossible, but none are wingless.

Tiny stingless bee species exist whose workers are less than 2 mm (0. 079 in) long. The largest bee in the world is *Megachile pluto*, a leafcutter bee whose females can attain a length of 39 mm (1. 5"). Members of the family Halictidae, or sweat bees, are the most common type of bee in the Northern Hemisphere, though they are small and often mistaken for wasps or flies.

The best-known bee species is the European honey bee, which, as its name suggests, produceshoney, as do a few other types of bee. Human management of this species is known as beekeeping or apiculture.

Bees are the favorite meal of *Merops apiaster*, the bee-eater (a bird). Other common predators arekingbirds, mockingbirds, beewolves, and dragonflies.

EVOLUTION

The ancestors of bees were wasps in the family Crabronidae, and therefore predators of other insects. The switch from insect prey to pollen may have resulted from the consumption of prey insects which were flower visitors and were partially covered with pollen when they were fed to the wasp larvae. This same evolutionary scenario may also have occurred within the vespoid wasps, where the group known as "pollen wasps" also evolved from predatory ancestors. Up until recently, the oldest non-compression bee fossil had been *Cretotrigona prisca* in New Jersey amber and of Cretaceous age, a meliponine. A recently reported bee fossil, of the genus *Melittosphex*, is considered *"an extinct lineage of pollen-collecting Apoidea sister to the modern bees"*, and dates from the early Cretaceous (~100 mya). Derived features of its morphology ("apomorphies") place it clearly within the bees, but it retains two unmodified ancestral traits ("plesiomorphies") of the legs (two mid-tibial spurs, and a slender hind basitarsus), indicative of its transitional status.

The earliest animal-pollinated flowers were pollinated by insects such as beetles, so the syndrome of insect pollination was well established before bees first appeared. The novelty is that bees are *specialized* as pollination agents, with behavioral and physical modifications that specifically enhance pollination, and are generally more efficient at the task than any other pollinating insect such as beetles, flies, butterfliesand pollen wasps. The appearance of such floral specialists is believed to have driven the adaptive radiation of the angiosperms, and, in turn, the bees themselves.

Among living bee groups, the "short-tongued" bee family Colletidae has traditionally been considered the most "primitive", and sister taxon to the remainder of the bees. In the 21st century, however, some researchers have claimed that the Dasypodaidae is the basal group, the short, wasp-like mouthparts of colletids being the result of convergent evolution, rather than indicative of a plesiomorphic condition. This subject is still under debate, and the phylogenetic relationships among bee families are poorly understood.

Eusocial and semisocial bees

Bees may be solitary or may live in various types of communities. The most advanced of these are eusocialcolonies found among the honey bees, bumblebees, and stingless bees. Sociality, of several different types, is believed to have evolved separately many times within the bees.

In some species, groups of cohabiting females may be sisters, and if there is a division of labor within the group, then they are considered semisocial.

Fig. 1. Honey bees defend against waspattacks

If, in addition to a division of labor, the group consists of a mother and her daughters, then the group is called eusocial. The mother is considered the queen and the daughters are workers. These castes may be purely behavioral alternatives, in which case the system is considered "primitively eusocial" (similar to many paper wasps), and if the castes are morphologically discrete, then the system is "highly eusocial".

There are many more species of primitively eusocial bees than highly eusocial bees, but they have rarely been studied. The biology of most such species is almost completely unknown. The vast majority are in the familyHalictidae, or "sweat bees". Colonies are typically small, with a dozen or fewer workers, on average. The only physical difference between queens and workers is average size, if they differ at all. Most species have a single season colony cycle, even in the tropics, and only mated females (future queens, or "gynes") hibernate (called diapause). A few species have long active seasons and attain colony sizes in the hundreds. The orchid bees include a number of primitively eusocial species with similar biology. Certain species of allodapine bees (relatives of carpenter bees) also have primitively eusocial colonies, with unusual levels of interaction between the adult bees and the developing brood. This is "progressive provisioning"; a larva's food is supplied gradually as it develops. This system is also seen in honey bees and some bumblebees.

Thanks to new molecular phylogeny data, a timeline of the evolution of eusocial bee species is available. The complexity of social behaviors seems positively correlated with age of evolutionary origin. This ranges from around 21 Mya in Halictidae, 53 Mya in allodapines, with the corbiculate Apidae (the group that includes honey bees and bumblebees) being the oldest at roughly 87 Mya.

Highly eusocial bees live in colonies. Each colony has a single queen, many workers and, at certain stages in the colony cycle, drones. When humans provide the nest, it is called a hive. Honey bee hives can contain up to 40, 000 bees at their annual peak, which occurs in the spring, but usually have fewer.

BUMBLEBEES

Bumblebees (*Bombus terrestris*, *Bombus pratorum*, et al.) are eusocial in a manner quite similar to the eusocial Vespidae such as hornets. The queen initiates a nest on her own (unlike queens of honey bees and stingless bees which start nests via swarms in the company of a large worker force). Bumblebee colonies typically have from 50 to 200 bees at peak population, which occurs in mid to late summer. Nest architecture is simple, limited by the size of the pre-existing nest cavity, and colonies are rarely perennial. Bumblebee queens sometimes seek winter safety in honey bee hives, where they are sometimes found dead in the spring by beekeepers, presumably stung to death by the honey bees. It is unknown whether any survive winter in such an environment.

Bumblebees are one of the more important wild pollinators, but have declined significantly in recent decades. In the UK, two species have become nationally extinct during the last 75 years while others have been placed on the UK Biodiversity Action Plan as priority species in recognition of the need for conservation action. In 2006 a new charity, the Bumblebee Conservation Trust, was established to coordinate efforts to conserve remaining populations through conservation and education. In 2011, the International Union for the Conservation of Natureset up the Bumblebee Specialist Group to review the threat status of all bumblebee species world-wide using the IUCN Red List criteria.

Stingless bees

Stingless bees are very diverse in behavior, but all are highly eusocial. They practise mass provisioning, complex nest architecture, and perennial colonies.

Honey bees

The true honey bees (genus *Apis*) have arguably the most complex social behavior among the bees, is the best known bee species, and one of the best known of all insects. There are 29 subspecies of*Apis mellifera*, native to Europe, the Middle East, and Africa.

Africanized honey bee

Fig. 2. An African bee Apis melliferasubspecies scutellata, extracts nectar from a flower as pollen grains stick to its body in Tanzania

Africanized bees, also called killer bees, are a hybrid strain of Apis mellifera derived from experiments by Warwick Estevam Kerr to cross European and African honey bee subspecies. Several queen bees escaped from his laboratory in South America and have spread throughout the Americas. Africanized honey bees are more defensive than European honey bees.

Solitary and communal bees

Most other bees, including familiar species of bee such as the Eastern carpenter bee (*Xylocopa virginica*), alfalfa leafcutter bee (*Megachile rotundata*), orchard mason bee (*Osmia lignaria*) and the hornfaced bee (*Osmia cornifrons*) are solitary in the sense that every female is fertile, and typically inhabits a nest she constructs herself.

There are no *worker* bees for these species. Solitary bees typically produce neither honey nor beeswax. They are immune from acarine and Varroa mites, but have their own unique parasites, pests and diseases.

Solitary bees are important pollinators, and pollen is gathered for provisioning the nest with food for their brood. Often it is mixed with nectar to form a paste-like consistency. Some solitary bees have very advanced types of pollen-carrying structures on their bodies.

A very few species of solitary bees are being increasingly cultured for commercial pollination. Most of these species belong to a distinct set of genera, namely:carpenter bees, sweat bees, mason bees, polyester bees, squash bees, dwarf carpenter bees, leafcutter bees, alkali bees, digger bees.

Solitary bees are often oligoleges, in that they only gather pollen from one or a few species/genera of plants (unlike honey bees and bumblebees which are generalists).

No known bees are nectar specialists; many oligolectic bees will visit multiple plants for nectar, but there are no bees which visit only one plant for nectar while also gathering pollen from many different sources. Specialist pollinators also include bee species which gather floral oils instead of pollen, and male orchid bees, which gather aromatic compounds from orchids (one of the only cases where male bees are effective pollinators).

In a very few cases only one species of bee can effectively pollinate a plant species, and some plants are endangered at least in part because their pollinator is dying off.

There is, however, a pronounced tendency for oligolectic bees to be associated with common, widespread plants which are visited by multiple pollinators (e. g., there are some 40 oligoleges associated with creosote bush in the US desert southwest, and a similar pattern is seen in sunflowers, asters, mesquite, etc.)

Solitary bees create nests in hollow reeds or twigs, holes in wood, or, most commonly, in tunnels in the ground. The female typically creates a compartment (a "cell") with an egg and some provisions for the resulting larva, then seals it off.

A nest may consist of numerous cells. When the nest is in wood, usually the last (those closer to the entrance) contain eggs that will become males. The adult does not provide care for the brood once the egg is laid, and usually dies after making one or more nests. The males typically emerge first and are ready for mating when the females emerge. Providing nest boxes for solitary bees is increasingly popular among gardeners. Solitary bees are either stingless or very unlikely to sting (only in self-defense, if ever).

While solitary females each make individual nests, some species are gregarious, preferring to make nests near others of the same species, giving the appearance to the casual observer that they are social. Large groups of solitary bee nests are called *aggregations*, to distinguish them from colonies.

In some species, multiple females share a common nest, but each makes and provisions her own cells independently. This type of group is called "communal" and is not uncommon. The primary advantage appears to be that a nest entrance is easier to defend from predators and parasites when there are multiple females using that same entrance on a regular basis.

Brood parasites

Brood parasites occur in several bee families including the apid subfamily Nomadinae. Females of these bees lack pollen collecting structures (the scopa) and do not construct their own nests. They typically enter the nests of pollen collecting species, and lay their eggs in cells provisioned by the host bee. When the cuckoo bee larva hatches it consumes the host larva's pollen ball, and kills and eats the host larva.

The cuckoo bees in the Bombus subgenus *Psithyrus* are closely related to, and resemble, their hosts in looks and size. This common pattern gave rise to the ecological principle "Emery's Rule". Others parasitize bees in different families, like Townsendiella, a nomadine apid, one species of which is a cleptoparasite of the dasypodaid genus Hesperapis, while the other species in the same genus attackhalictid bees.

Nocturnal bees

Four bee families (Andrenidae, Colletidae, Halictidae, and Apidae) contain some species that arecrepuscular (these may be either the vespertine or matinal type). These bees have greatly enlarged ocelli, which are extremely sensitive to light and dark, though incapable of forming images. Many are pollinators of flowers that themselves are crepuscular, such as evening primroses, and some live in desert habitats where daytime temperatures are extremely high.

Flight

In Antoine Magnan's 1934 book *Le vol des insectes*, he wrote that he and André Sainte-Laguë had applied the equations of air resistance to insects and found that their flight could not be explained by fixed-wing calculations, but that "One shouldn't be surprised that the results of the calculations don't square with reality". This has led to a common misconception that bees "violate aerodynamic theory", but in fact it merely confirms that bees do not engage in fixed-wing flight, and that their flight is explained by other mechanics, such as those used by helicopters.

In 1996, Charlie Ellington at Cambridge University showed that vortices created by many insects' wings and non-linear effects were a vital source of lift; vortices and non-linear phenomena are notoriously difficult areas of hydrodynamics, which has made for slow progress in theoretical understanding of insect flight. In 2005, Michael Dickinson and his Caltech colleagues studied honey bee flight with the assistance of high-speed cinematography and a giant robotic mock-up of a bee wing. Their analysis revealed that sufficient lift was generated by "the unconventional combination of short, choppy wing strokes, a rapid rotation of the wing as it flops over and reverses direction, and a very fast wing-beat frequency". Wing-beat frequency normally increases as size decreases, but as the bee's wing beat covers such a small arc, it flaps approximately 230 times per second, faster than a fruitfly (200 times per second) which is 80 times smaller.

Pollination

Bees play an important role in pollinating flowering plants, and are the major type of pollinator inecosystems that contain flowering plants. Bees either focus on gathering nectar or on gathering pollen depending on demand, especially in social species. Bees gathering nectar may accomplish pollination, but bees that are deliberately gathering pollen are more efficient pollinators. It is estimated that one third of the human food supply depends on insect pollination, most of which is accomplished by bees, especially the domesticated European honey bee. Contract pollination has overtaken the role of honey production for beekeepers in many countries. Monoculture and the massive decline of many bee species (both wild and domesticated) have increasingly caused honey bee keepers to become migratory so that bees can be concentrated in seasonally varying high-demand areas of pollination.

Most bees are fuzzy and carry an electrostatic charge, which aids in the adherence of pollen. Female bees periodically stop foraging and groom themselves to pack the pollen into the scopa, which is on the legs in most bees, and on the ventral abdomen on others, and modified into specialized pollen baskets on the legs of honey bees and their relatives.

Many bees are opportunistic foragers, and will gather pollen from a variety of plants, while others are oligolectic, gathering pollen from only one or a few types of plant. A small number of plants produce nutritious floral oils rather than pollen, which are gathered and used by oligolectic bees. One small subgroup of stingless bees, called "vulture bees, " is specialized to feed on carrion, and these are the only bees that do not use plant products as food.

Pollen and nectar are usually combined to form a "provision mass", which is often soupy, but can be firm. It is formed into various shapes (typically spheroid), and stored in a small chamber (a "cell"), with the egg deposited on the mass. The cell is typically sealed after the egg is laid, and the adult and larva never interact directly (a system called "mass provisioning").

In New Zealand scientists discovered that three genera of native bees have evolved to open flower buds of the native mistletoe *Peraxilla tetrapetala*. The buds cannot open themselves but are visited by birds such as the tui and bellbird which twist the top of the ripe bud. That action releases a mechanism which causes the petals to suddenly spring open, giving access to the nectar and pollen.

However, when observing the native bees in the Canterbury province in the South Island, the scientists were astonished to see the bees biting the top off the buds, then pushing with their legs, occasionally popping open the buds to allow the bees to harvest the nectar and pollen, and therefore aid in the pollination of the mistletoe which is in decline in New Zealand. Nowhere else in the world have bees demonstrated ability to open explosive bird-adapted flowers.

Visiting flowers can be a dangerous occupation. Many assassin bugs and crab spiders hide in flowers to capture unwary bees. Other bees are lost to birds in flight. Insecticides used on blooming plants kill many bees, both by direct poisoning and by contamination of their food supply.

A honey bee queenmay lay 2000 eggs per day during spring buildup, but she also must lay 1000 to 1500 eggs per day during the foraging season, mostly to replace daily casualties, most of which are workers dying of old age. Among solitary and primitively social bees, however, lifetime reproduction is among the lowest of all insects, as it is common for females of such species to produce fewer than 25 offspring.

The population value of bees depends partly on the individual efficiency of the bees, but also on the population itself. Thus while bumblebeeshave been found to be about ten times more efficient pollinators on cucurbits, the total efficiency of a colony of honey bees is much greater due to greater numbers. Likewise during early spring orchard blossoms, bumblebee populations are limited to only a few queens, and thus are not significant pollinators of early fruit.

Bees and humans

Bees figure prominently in mythology and folklore and have been used by political theorists as a model for human society. Journalist Bee Wilson states that the image of a community of honey bees "occurs from ancient to modern times, in Aristotle and Plato; in Virgil and Seneca; in Erasmus andShakespeare; Tolstoy, as well as by social theorists Bernard Mandeville and Karl Marx. " They are found in heraldry where they can signify industriousness as in the Manchester bee in the crest ofManchester City Council.

In English folklore, bees would be told of important events in the household, in a custom known as "Telling the bees".

Despite the honey bee's painful sting and the stereotype of insects as pests, bees are generally held in high regard. This is most likely due to their usefulness as pollinators and as producers of honey, their social nature, and their reputation for diligence.

Bees are one of the few insects frequently used in advertisements in a positive manner, typically for products containing honey (such as Honey Nut Cheerios).

In ancient Egypt, the bee was seen to symbolize the lands of Lower Egypt, with the Pharaoh being referred to as "He of Sedge and Bee" (the sedge representing Upper Egypt).

In North America, yellowjackets and hornets, especially when encountered as flying pests, are often misidentified as bees, despite numerous differences between them.

Although a bee sting can be deadly to those with allergies, virtually all bee species are non-aggressive if undisturbed and many cannot sting at all. Humans are often a greater danger to bees, as bees can be affected or even harmed by encounters with toxic chemicals in the environment.

In Indonesia bee larvae are eaten as a companion to rice, after being mixed with shredded coconut "meat", wrapped in banana leaves, and steamed.

Pollinator decline

From 1972 to 2006, there was a dramatic reduction in the number of feral honey bees in the US, which are now almost absent. At the same time there was a significant though somewhat gradual decline in the number of colonies maintained by beekeepers.

This decline includes the cumulative losses from all factors, such as urbanization, systematic pesticide use, tracheal and Varroa mites, and commercial beekeepers' retiring and going out of business. However, in late 2006 and early 2007 the rate of attrition reached new proportions, and the term colony collapse disorder was coined to describe the sudden disappearances.

After several years of research and concern, a team of scientists headed by Jerry Bromenshenk published a paper in October 2010 saying that a new DNA-based virus, invertebrate iridescent virus or IIV6, and the fungus Nosema ceranae were found in every killed colony the group studied.

In their study they found that neither agent alone seemed deadly, but a combination of the virus and *Nosema ceraneae* was always 100% fatal. Bromenshenk said it is not yet clear whether one condition weakens the bees enough to be finished off by the second, or whether they somehow compound the other's destructive power.

"They're co-factors, that's all we can say at the moment. They're both present in all these collapsed colonies. " Investigations into the phenomenon had occurred amidst great concern over the nature and extent of the losses.

In 2009 some reports from the US suggested that 1/3 of the honey bee colonies did not survive the winter, though normal winter losses are known to be around 25%. At the end of May 2012, the Swiss government reported that about half of the bee population had not survived the winter. The main cause of the decline was thought to be the parasite varroa.

Apart from colony collapse disorder, many of the losses outside the US have also been attributed to other causes. Pesticides used to treat seeds, such as Clothianidin, Imidacloprid and Thiamethoxam, have been considered prime suspects. Other species of bees such asmason bees are increasingly cultured and used to meet the agricultural pollination need.

Native pollinators include bumblebees and solitary bees, which often survive in refuges in wild areas away from agricultural spraying, but may still be poisoned in massive spray programs for mosquitoes, gypsy moths, or other insect pests. Although pesticide use remains a concern, the major problem for wild pollinator populations is the loss of the flower-rich habitat on which they depend for food. Throughout the northern hemisphere, the last 70 or so years have seen an intensification of agricultural systems, which has decreased the abundance and diversity of wild flowers.

Legislation such as the UK's Bees Act 1980 is designed to stop the decline of bees. In April 2013 the European Union announced plans to restrict the use of certain pesticides to stop bee populations from declining further.

In 2014 the Intergovernmental Panel on Climate Change report warned that bees and other pollinators faced increased risk of extinction because of global warming. This was due to their natural habitats moving and the alteration in the seasonal behaviour of species. Climate change was causing the natural foraging areas for bees to change location but the bees were not always able to move to these new areas. Bees were also emerging at different times in the year due to global warming when flowering plants were not available.

HONEY

Honey is a sweet food made by bees using nectar from flowers. The variety produced byhoney bees (the genus *Apis*) is the one most commonly referred to, as it is the type of honey collected by most beekeepers and consumed by people. Honeys are also produced bybumblebees, stingless bees, and other hymenopteran insects such as honey wasps, though the quantity is generally lower and they have slightly different properties compared to honey from the genus *Apis*.

Honey bees convert nectar into honey by a process of regurgitation and evaporation. They store it as a primary food source in wax honeycombs inside the beehive.

Honey gets its sweetness from the monosaccharides fructose and glucose, and has about the same relative sweetness as granulated sugar. It has attractive chemical properties for baking and a distinctive flavor that leads some people to prefer it over sugar and other sweeteners.

Most microorganisms do not grow in honey because of its low water activity of 0. 6. However, honey sometimes contains dormant endospores of the bacterium Clostridium botulinum, which can be dangerous to infants, as the endospores can transform into toxin-producing bacteria in infants' immature intestinal tracts, leading to illness and even death.

Honey has had a long history in human consumption, and is used in various foods and beverages as a sweetener and flavoring. It also has a role in religion and symbolism. Flavors of honey vary based on the nectar source, and various types and grades of honey are available.

It has also been used in various medicinal traditions to treat ailments. The study of pollens and spores in honey (melissopalynology) can determine floral sources of honey.

Bees carry an electrostatic charge whereby they attract other particles in addition to pollen, which become incorporated into their honey; honey can be analysed by the technique of melissopalynology, in the area of environmental studies of radioactive particles, dust, and particulate pollution.

Formation

Honey's natural sugars are dehydrated, which prevents fermentation, with added enzymes to modify and transform their chemical composition and pH. Invertases and digestive acids hydrolyze sucrose to give the monosaccharides glucose and fructose. Invertase is one of these enzymes synthesized by the body of the insect.

Honey bees transform saccharides into honey by a process of regurgitation, a number of times, until it is partially digested. The bees do the regurgitation and digestion as a group. After the last regurgitation, the aqueous solution is still high in water, so the process continues by evaporation of much of the water and enzymatic transformation.

Honey is produced by bees as a food source. To produce about 500 g of honey, foraging honey bees have to travel the equivalent of three times around the world. In cold weather or when fresh food sources are scarce, bees use their stored honey as their source of energy. By contriving for bee swarms to nest in artificial hives, people have been able to semidomesticate the insects, and harvest excess honey. In the hive (or in a wild nest), the three types of bees are:

- a single female queen bee
- a seasonally variable number of male drone bees to fertilize new queens
- some 20, 000 to 40, 000 female worker bees

The worker bees raise larvae and collect the nectar that will become honey in the hive. Leaving the hive, they collect sugar-rich flower nectarand return.

In the hive, the bees use their "honey stomachs" to ingest and regurgitate the nectar a number of times until it is partially digested. Invertase synthesized by the bees and digestive acids hydrolyze sucrose to give the same mixture of glucose and fructose.

The bees work together as a group with the regurgitation and digestion until the product reaches a desired quality. It is then stored in honeycomb cells. After the final regurgitation, the honeycomb is left unsealed. However, the nectar is still high in both water content and natural yeasts, which, unchecked, would cause the sugars in the nectar to ferment.

The process continues as bees inside the hive fan their wings, creating a strong draft across the honeycomb, which enhances evaporation of much of the water from the nectar. This reduction in water content raises the sugar concentration and prevents fermentation. Ripe honey, as removed from the hive by a beekeeper, has a long shelf life, and will not ferment if properly sealed.

Another source of honey is from a number of wasp species, such as the wasps Brachygastra lecheguana and Brachygastra mellifica, which are found in South and Central America. These species are known to feed on nectar and produce honey.

Some wasps, such as the Polistes versicolor, even consume honey themselves, switching from feeding on pollen in the middle of their lifecycles to feeding on honey, which can better provide for their energy needs.

Collection

Honey is collected from wild bee colonies, or from domesticated beehives. Wild bee nests are sometimes located by following a honeyguidebird. The bees may first be pacified by using smoke from a bee smoker. The smoke triggers a feeding instinct (an attempt to save the resources of the hive from a possible fire), making them less aggressive and the smoke obscures the pheromones the bees use to communicate.

The honeycomb is removed from the hive and the honey may be extracted from that, either by crushing or by using a honey extractor. The honey is then usually filtered to remove beeswax and other debris.

Before the invention of removable frames, bee colonies were often sacrificed in order to conduct the harvest. The harvester would take all the available honey and replace the entire colony the next spring.

Since the invention of removable frames, the principles of husbandry lead most beekeepers to ensure that their bees will have enough stores to survive the winter, either by leaving some honey in the beehive or by providing the colony with a honey substitute such as sugar water or crystalline sugar (often in the form of a "candyboard"). The amount of food necessary to survive the winter depends on the race of bees and on the length and severity of local winters.

Modern uses

As a food and in cooking

Over its history as a food, the main uses of honey are in cooking, baking, desserts, as a spread on bread, and as an addition to various beverages, such as tea, and as a sweetener in some commercial beverages. Honey barbecue and honey mustard are common and popular sauce flavors.

As a fermentable

Honey is the main ingredient in the alcoholic beverage mead, which is also known as "honey wine" or "honey beer". Historically, the ferment for mead was honey's naturally occurring yeast. Honey is also used as an adjunct in some beers.

Honey wine, or mead, is typically (modern era) made with a honey and water mixture with yeast added for fermentation. Primary fermentation usually takes 40 days, after which the must needs to be racked into a secondary fermentation vessel and left to sit about 35–40 more days.

If done properly, fermentation will be finished by this point (though if a sparkling mead is desired, fermentation can be restarted after bottling by the addition of a small amount of sugar), but most meads require aging for 6–9 months or more in order to be palatable.

Honey-producing and consuming countries

Mexico is also an important producer of honey, providing more than 4% of the world's supply. Much of this (about one-third) comes from theYucatán Peninsula. Honey production began there when the *Apis melliferaa*nd the *A. mellifera ligustica* were introduced there early in the 20th century.

Most of Mexico's Yucatán producers are small, family operations who use original traditional techniques, moving hives to take advantage of the various tropical and subtropical flowers.

Honey is also one of the gourmet products of the French island of Corsica. Corsican honey is certified as to its origin (Appellation d'origine contrôlée) just as are French wines, like Champagne.

Honey consumption per capita per year exceeds one kilogram in some countries like Austria, Germany and Switzerland.

Physical and chemical properties

The physical properties of honey vary, depending on water content, the type of flora used to produce it (pasturage), temperature, and the proportion of the specific sugars it contains. Fresh honey is asupersaturated liquid, containing more sugar than the water can typically dissolve at ambient temperatures.

At room temperature, honey is a supercooled liquid, in which the glucose will precipitate into solid granules. This forms a semisolid solution of precipitated glucose crystals in a solution offructose and other ingredients.

Phase transitions

The melting point of crystallized honey is between 40 and 50 °C (104 and 122 °F), depending on its composition. Below this temperature, honey can be either in a metastable state, meaning that it will not crystallize until a seed crystal is added, or, more often, it is in a "labile" state, being saturated with enough sugars to crystallize spontaneously.

The rate of crystallization is affected by many factors, but the primary factor is the ratio of the main sugars: fructose to glucose. Honeys that are supersaturated with a very high percentage of glucose, such as brassica honey, will crystallize almost immediately after harvesting, while honeys with a low percentage of glucose, such as chestnut or tupelo honey, do not crystallize. Some types of honey may produce very large but few crystals, while others will produce many small crystals.

Crystallization is also affected by water content, because a high percentage of water will inhibit crystallization, as will a high dextrin content. Temperature also affects the rate of crystallization, with the fastest growth occurring between 13 and 17 °C (55 and 63 °F). Crystal nuclei tend to form more readily if the honey is disturbed, by stirring, shaking or agitating, rather than if left at rest. However, the nucleationof microscopic seed-crystals is greatest between 5 and 8 °C (41 and 46 °F). Therefore, larger but fewer crystals tend to form at higher temperatures, while smaller but more-numerous crystals usually form at lower temperatures. Below 5 °C, the honey will not crystallize and, thus, the original texture and flavor can be preserved indefinitely.

Since honey normally exists below its melting point, it is a supercooled liquid. At very low temperatures, honey will not freeze solid. Instead, as the temperatures become lower, the viscosity of honey increases. Like most viscous liquids, the honey will become thick and sluggish with decreasing temperature. At "20 °C ("4 °F), honey may appear or even feel solid, but it will continue to flow at very low rates. Honey has aglass transition between "42 and "51 °C ("44 and "60 °F). Below this temperature, honey enters a glassy state and will become anamorphous solid (noncrystalline).

Viscosity

The viscosity of honey is affected greatly by both temperature and water content. The higher the water percentage, the easier honey flows. Above its melting point, however, water has little effect on viscosity. Aside from water content, the composition of honey also has little effect on viscosity, with the exception of a few types. At 25 °C (77 °F), honey with 14% water content generally has a viscosity around 400 poise, while a honey containing 20% water has a viscosity around 20 poise. Viscosity increase due to temperature occurs very slowly at first. A honey containing

16% water, at 70 °C (158 °F), will have a viscosity around 2 poise, while at 30 °C (86 °F), the viscosity is around 70 poise. As cooling progresses, honey becomes more viscous at an increasingly rapid rate, reaching 600 poise around 14 °C (57 °F). However, while honey is very viscous, it has rather low surface tension.

A few types of honey have unusual viscous properties. Honeys from heather or manuka display thixotropic properties. These types of honey enter a gel-like state when motionless, but then liquify when stirred.

Electrical and optical properties

Because honey contains electrolytes, in the form of acids and minerals, it exhibits varying degrees of electrical conductivity. Measurements of the electrical conductivity are used to determine the quality of honey in terms of ash content.

The effect honey has on light is useful for determining the type and quality. Variations in the water content alter the refractive index of honey. Water content can easily be measured with a refractometer. Typically, the refractive index for honey will range from 1. 504 at 13% water content to 1. 474 at 25%. Honey also has an effect on polarized light, in that it will rotate the polarization plane.

The fructose will give a negative rotation, while the glucose will give a positive one. The overall rotation can be used to measure the ratio of the mixture. Honey may vary in color between pale yellow and dark brown, but other bright colors may occasionally be found, depending on the source of the sugar harvested by the bees.

Hygroscopy and fermentation

Honey has the ability to absorb moisture directly from the air, a phenomenon called hygroscopy. The amount of water the honey will absorb is dependent on the relative humidity of the air.

Because honey contains yeast, this hygroscopic nature requires that honey be stored in sealed containers to prevent fermentation, which usually begins if the honey's water content rises much above 25%. Honey will tend to absorb more water in this manner than the individual sugars would allow on their own, which may be due to other ingredients it contains.

Fermentation of honey will usually occur after crystallization because, without the glucose, the liquid portion of the honey primarily consists of a concentrated mixture of the fructose, acids, and water, providing the yeast with enough of an increase in the water percentage for growth.

Honey that is to be stored at room temperature for long periods of time is often pasteurized, to kill any yeast, by heating it above 70 °C (158 °F).

Thermal characteristics

Like all sugar compounds, honey will caramelize if heated sufficiently, becoming darker in color, and eventually burn. However, honey contains fructose, which caramelizes at lower temperatures than the glucose.

The temperature at which caramelization begins varies, depending on the composition, but is typically between 70 and 110 °C (158 and 230 °F). Honey also contains acids, which act as catalysts, decreasing the caramelization temperature even more.Of these acids, the amino acids, which occur in very small amounts, play an important role in the darkening of honey.

The amino acids form darkened compounds called melanoidins, during a Maillard reaction. The Maillard reaction will occur slowly at room temperature, taking from a few to several months to show visible darkening, but will speed-up dramatically with increasing temperatures. However, the reaction can also be slowed by storing the honey at colder temperatures.

Unlike many other liquids, honey has very poor thermal conductivity, taking a long time to reach thermal equilibrium. Melting crystallized honey can easily result in localized caramelization if the heat source is too hot, or if it is not evenly distributed.

However, honey will take substantially longer to liquify when just above the melting point than it will at elevated temperatures. Melting 20 kilograms of crystallized honey, at 40 °C (104 °F), can take up to 24 hours, while 50 kilograms may take twice as long.

These times can be cut nearly in half by heating at 50 °C (122 °F). However, many of the minor substances in honey can be affected greatly by heating, changing the flavor, aroma, or other properties, so heating is usually done at the lowest temperature possible for the shortest amount of time.

Classification

Honey is classified by its floral source, and there are also divisions according to the packaging and processing used. There are also regional honeys. In the USA honey is also graded on its color and optical density by USDA standards, graded on the Pfund scale, which ranges from 0 for "water white" honey to more than 114 for "dark amber" honey.

Floral source

Generally, honey is classified by the floral source of the nectar from which it was made. Honeys can be from specific types of flower nectars or can be blended after collection. The pollen in honey is traceable to floral source and therefore region of origin. The rheological &melissopalynological properties of honey can be used to identify the major plant nectar source used in its production.

Blended

Most commercially available honey is blended, meaning it is a mixture of two or more honeys differing in floral source, color, flavor, density or geographic origin.

Polyfloral

Polyfloral honey, also known as wildflower honey, is derived from the nectar of many types of flowers.

The taste may vary from year to year, and the aroma and the flavor can be more or less intense, depending on which bloomings are prevalent.

Monofloral

Monofloral honey is made primarily from the nectar of one type of flower. Different monofloral honeys have a distinctive flavor and color because of differences between their principal nectar sources.

To produce monofloral honey, beekeepers keep beehives in an area where the bees have access to only one type of flower. In practice, because of the difficulties in containing bees, a small proportion of any honey will be from additional nectar from other flower types.

Typical examples of North American monofloral honeys are clover, orange blossom, blueberry, sage, tupelo, buckwheat, fireweed, mesquite and sourwood. Some typical European examples include thyme, thistle, heather, acacia, dandelion, sunflower, honeysuckle, and varieties from lime and chestnut trees. In North Africa (e. g. Egypt) examples include clover, cotton, and citrus (mainly orange blossoms).

Honeydew honey

Instead of taking nectar, bees can take honeydew, the sweet secretions of aphids or other plant sap-sucking insects. Honeydew honey is very dark brown in color, with a rich fragrance of stewed fruit or fig jam, and is not as sweet as nectar honeys. Germany's Black Forest is a well known source of honeydew-based honeys, as well as some regions in Bulgaria, Tara (mountain) in Serbia and Northern California in the United States. In Greece, pine honey (a type of honeydew honey) constitutes 60–65% of the annual honey production. Honeydew honey is popular in some areas, but in other areas beekeepers have difficulty selling the stronger flavored product.

The production of honeydew honey has some complications and dangers. The honey has a much larger proportion of indigestibles than light floral honeys, thus causing dysentery to the bees, resulting in the death of colonies in areas with cold winters. Good beekeeping management requires the removal of honeydew prior to winter in colder areas. Bees collecting this resource also have to be fed protein supplements, as honeydew lacks the protein-rich pollen accompaniment gathered from flowers.

Classification by packaging and processing

Generally, honey is bottled in its familiar liquid form. However, honey is sold in other forms, and can be subjected to a variety of processing methods.

- **Crystallized honey** is honey in which some of the glucose content has spontaneously crystallized from solution as the monohydrate. Also called "granulated honey" or "candied honey. " Honey that has crystallized (or commercially purchased crystallized) can be returned to a liquid state by warming.
- **Pasteurized honey** is honey that has been heated in a pasteurization process which requires temperatures of 161 °F (72 °C) or higher. Pasteurization destroys yeast cells. It also liquefies any microcrystals in the honey, which delays the onset of visible crystallization. However, excessive heat exposure also results in product deterioration, as it

increases the level of hydroxymethylfurfural(HMF) and reduces enzyme (e. g. diastase) activity. Heat also affects appearance (darkens the natural honey color), taste, and fragrance.

- **Raw honey** is honey as it exists in the beehive or as obtained by extraction, settling or straining, without adding heat (although some honey that has been "minimally processed" is often labeled as raw honey). Raw honey contains some pollen and may contain small particles of wax. Someallergy sufferers wrongly believe that raw, local honey can help build tolerance to the pollen in the air.
- **Strained honey** has been passed through a mesh material to remove particulate material (pieces of wax, propolis, other defects) without removing pollen, minerals or enzymes.
- **Filtered honey** is honey of any type that has been filtered to the extent that all or most of the fine particles, pollen grains, air bubbles, or other materials normally found in suspension, have been removed. The process typically heats honey to 150–170 °F (66–77 °C) to more easily pass through the filter. Filtered honey is very clear and will not crystallize as quickly, making it preferred by the supermarket trade.
- **Ultrasonicated honey** has been processed by ultrasonication, a non-thermal processing alternative for honey. When honey is exposed to ultrasonication, most of the yeast cells are destroyed. Those cells that survive sonication generally lose their ability to grow, which reduces the rate of honey fermentation substantially. Ultrasonication also eliminates existing crystals and inhibits further crystallization in honey. Ultrasonically aided liquefaction can work at substantially lower temperatures of approximately 95 °F (35 °C) and can reduce liquefaction time to less than 30 seconds.
- **Creamed honey**, also called whipped honey, spun honey, churned honey, honey fondant, and (in the UK) set honey, has been processed to control crystallization. Creamed honey contains a large number of small crystals, which prevent the formation of larger crystals that can occur in unprocessed honey. The processing also produces a honey with a smooth, spreadable consistency.
- **Dried honey** has the moisture extracted from liquid honey to create completely solid, nonsticky granules. This process may or may not include the use of drying and anticaking agents. Dried honey is used in baked goods, and to garnish desserts.
- **Comb honey** is honey still in the honeybees' wax comb. It traditionally is collected by using standard wooden frames in honey supers. The

frames are collected and the comb is cut out in chunks before packaging. As an alternative to this labor-intensive method, plastic rings or cartridges can be used that do not require manual cutting of the comb, and speed packaging. Comb honey harvested in the traditional manner is also referred to as "cut-comb honey".

- **Chunk honey** is packed in widemouth containers consisting of one or more pieces of comb honey immersed in extracted liquid honey.
- **Honey decoctions** are made from honey or honey by-products which have been dissolved in water, then reduced (usually by means of boiling). Other ingredients may then be added. (For example, abbamele has added citrus.) The resulting product may be similar to molasses.

Preservation

Because of its unique composition and chemical properties, honey is suitable for long-term storage, and is easily assimilated even after long preservation. Honey, and objects immersed in honey, have been preserved for centuries. The key to preser-vation is limiting access to humidity. In its cured state, honey has a sufficiently high sugar content to inhibit fermentation. If exposed to moist air, itshydrophilic properties will pull moisture into the honey, eventually diluting it to the point that fermentation can begin.

Regardless of preservation, honey may crystallize over time. The crystals can be dissolved by heating the honey.

DISTINGUISHING

Grading

In the US, honey grading is performed voluntarily (USDA does offer inspection and grading "as on-line (in-plant) or lot inspection... upon application, on a fee-for-service basis. ") based upon USDA standards. Honey is graded based upon a number of factors, including water content, flavor and aroma, absence of defects and clarity. Honey is also classified by color though it is not a factor in the grading scale. The honey grade scale is:

Other countries may have differing standards on the grading of honey. India, for example, certifies honey grades based on additional factors, such as the Fiehe's test, and other empirical measurements.

Indicators of quality

High-quality honey can be distinguished by fragrance, taste, and consistency. Ripe, freshly collected, high-quality honey at 20 °C (68 °F) should flow from a knife in a straight stream, without breaking into separate drops.

After falling down, the honey should form a bead. The honey, when poured, should form small, temporary layers that disappear fairly quickly, indicating high viscosity. If not, it indicates excessive water content (over 20%) of the product. Honey with excessive water content is not suitable for long-term preservation.

In jars, fresh honey should appear as a pure, consistent fluid, and should not set in layers. Within a few weeks to a few months of extraction, many varieties of honey crystallize into a cream-colored solid. Some varieties of honey, including tupelo, acacia, and sage, crystallize less regularly.

Honey may be heated during bottling at temperatures of 40–49 °C (104–120 °F) to delay or inhibit crystallization. Overheating is indicated by change in enzyme levels, for instance, diastase activity, which can be determined with the Schade or the Phadebas methods. A fluffy film on the surface of the honey (like a white foam), or marble-colored or white-spotted crystallization on a containers sides, is formed by air bubbles trapped during the bottling process.

A 2008 Italian study determined nuclear magnetic resonance spectroscopy can be used to distinguish between different honey types, and can be used to pinpoint the area where it was produced. Researchers were able to identify differences in acacia and polyfloral honeys by the differing proportions of fructose and sucrose, as well as differing levels of aromatic amino acids phenylalanine and tyrosine. This ability allows greater ease of selecting compatible stocks.

Acid content and flavor effects

The average pH of honey is 3. 9, but can range from 3. 4 to 6. 1. Honey contains many kinds of acids, both organic and amino. However, the different types and their amounts vary considerably, depending on the type of honey. These acids may be aromatic or aliphatic (non-aromatic). The aliphatic acids contribute greatly to the flavor of honey by interacting with the flavors of other ingredients.

Organic acids comprise most of the acids in honey, accounting for 0. 17–1. 17% of the mixture, with gluconic acid formed by the actions of an enzyme called glucose oxidase as the most prevalent. Other organic acids are minor, consisting of formic, acetic, butyric, citric, lactic, malic, pyroglutamic, propionic, valeric, capronic, palmitic, and succinic, among many others.

Nutrition and composition

The specific composition, color, aroma and flavor of any batch of honey depend on the flowers foraged by bees that produced the honey. Honey, a mixture of sugars and other compounds, is mainly fructose (about 38-55%) and glucose (about 31%), with remaining carbohydrates including maltose, sucrose, and other complex carbohydrates. Its glycemic index ranges from 31 to 78, depending on the variety. Honey contains only trace amounts of vitamins or minerals(table).

Typical honey analysis:

- Fructose: 38. 2%
- Glucose: 31. 3%
- Maltose: 7. 1%
- Sucrose: 1. 3%
- Water: 17. 2%
- Higher sugars: 1. 5%
- Ash: 0. 2%
- Other/undetermined: 3. 2%

Honey has a density of about 1. 36 kilograms per litre (36% denser than water).

Adulteration

Adulteration of honey is the addition of other sugars, syrups or compounds into honey to change its flavor, viscosity, make it cheaper to produce, or to increase the fructose content in order to stave off crystallization.

According to the Codex Alimentarius of the United Nations, any product labeled as honey or pure honey must be a wholly natural product, although different nations have their own laws concerning labeling.

Adulteration of honey is sometimes used as a method of deception when buyers are led to believe that the honey is pure. The practice was common dating back to ancient times, when crystallized honey was often mixed with flour or other fillers, hiding the adulteration from buyers until the honey was liquefied. In modern times the most common adulteration-ingredient became clear, almost-flavorless corn syrup, which, when mixed with honey, is often very difficult to distinguish from unadulterated honey.

Isotope ratio mass spectrometry can be used to detect addition of corn syrup and cane sugar by the carbon isotopic signature. Addition of sugars originating from corn or sugar cane (C4 plants, unlike the plants used by bees, and also sugar beet, which are predominantly C3 plants) skews the isotopic ratio of sugars present in honey, but does not influence the isotopic ratio of proteins.

In an unadulterated honey, the carbon isotopic ratios of sugars and proteins should match. Levels as low as 7% of addition can be detected. In one country, the USA, according to The National Honey Board (a USDA-overseen organization), "honey stipulates a pure product that does not allow for the addition of any other substance... this includes, but is not limited to, water or other sweeteners".

Health applications

In myths and folk medicine, honey has been used by humans both orally and topically to treat various ailments including gastric disturbances, ulcers, wounds, and burns. For example, honey was used medicinally by ancient Greeks, Egyptians and in Ayurveda and traditional Chinese medicine. The Quran mentions honey as a medicine. Medieval Islamic and Christian scholars described the use of honey in medicine of that period.

Due mainly to low numbers and quality of human clinical trials, little to no efficacy has been identified about the potential health benefits of honey.

Allergies

Consumption is sometimes advocated as a treatment for seasonal allergies due to pollen. Honey is generally considered ineffective for the treatment of rhinosinusitis.

Skin wounds and burns

There is some weak evidence that honey may help treat skin wounds or mild burns if used in a dressing, but can delay healing in more severe burns. Using honey with burn dressings is not recommended.

Cancer

No evidence shows benefit of honey in treating cancer.

Cough

Little evidence supports honey as treatment of coughs for children. For chronic cough and acute cough, Cochrane systematic reviews found no strong evidence for or against the use of honey. US and UK regulatory authorities recommend avoiding giving over the counter cold medication to children. UK authorities recommend "a warm drink of lemon and honey or a simple cough syrup that contains glycerol or honey" in children instead, but warn that honey should not be given to very young children because of the risk of infant botulism.

Immune system

People who have a weakened immune system should not eat honey because of the risk of bacterial or fungal infection.

Ingrown toenail

Honey is of no benefit in treating ingrown toenail.

Ulcers

No evidence supports the use of honey-based products for the treatment of venous leg ulcers.

Health hazards

Botulism

Infants can contract botulism from honey.

Infantile botulism shows geographical variation. In the UK, only six cases have been reported between 1976 and 2006, yet the U. S. has much higher rates: 1. 9 per 100, 000 live births, 47. 2% of which are in California. While the risk honey poses to infant health is small, it is recommended not to take the risk.

Toxic honey

Honey produced from flowers of rhododendrons, mountain laurels, sheep laurel, and azaleas may cause honey intoxication. Symptoms include dizziness, weakness, excessive perspiration, nausea, and vomiting. Less commonly, low blood pressure, shock, heart rhythm irregularities, and convulsions may occur, with rare cases resulting in death. Honey intoxication is more likely when using "natural" unprocessed honey and honey from farmers who may have a small number of hives. Commercial processing, with pooling of honey from numerous sources, is thought to dilute any toxins.

Toxic honey may also result when bees are proximate to tutu bushes (*Coriaria arborea*) and the vine hopper insect (Scolypopa australis). Both are found throughout New Zealand. Bees gather honeydew produced by the vine hopper insects feeding on the tutu plant. This introduces the poison tutin into honey. Only a few areas in New Zealand (Coromandel Peninsula, Eastern Bay of Plenty and the Marlborough Sound) frequently produce toxic honey. Symptoms of tutin poisoning include vomiting, delirium, giddiness, increased excitability, stupor, coma, and violent convulsions. To reduce the risk of tutin poisoning, humans should not eat honey taken from feral hives in the risk areas of New Zealand. Since December 2001, New Zealand beekeepers have been required to reduce the risk of producing toxic honey by closely monitoring tutu, vine hopper, and foraging conditions within 3 kilometres (1. 9 mi) of their apiary.

IN HISTORY, CULTURE, AND FOLKLORE

Honey use and production has a long and varied history. In many cultures, honey has associations that go beyond its use as a food. Honey is frequently used as a talisman and symbol of sweetness.

Ancient times

Honey collection is an ancient activity. Humans apparently began hunting for honey at least 8, 000 years ago, as evidenced by a cave painting in Valencia, Spain. The painting is a Mesolithic rock painting, showing two honey-hunters collecting honey and honeycomb from a wild bee nest. The figures are depicted carrying baskets or gourds, and using a ladder or series of ropes to reach the wild nest.

The greater honeyguide bird guides humans to wild bee hives and this behavior may have evolved with early hominids

So far, the oldest remains of honey have been found in the country of Georgia. Archaeologists have found honey remains on the inner surface of clay vessels unearthed in an ancient tomb, dating back some 4, 700–5,

500 years. In ancient Georgia, honey was packed for people's journeys into the afterlife, and more than one type, too – along for the trip were linden, berry, and a meadow-flower variety.

In ancient Egypt, honey was used to sweeten cakes and biscuits, and was used in many other dishes. Ancient Egyptian and Middle Eastern peoples also used honey for embalming the dead. The fertility god of Egypt, Min, was offered honey.

The spiritual and therapeutic use of honey in ancient India is documented in both the Vedas and the Ayurvedatexts, which were both composed at least 4, 000 years ago.

Pliny the Elder devotes considerable space in his book Naturalis Historia to the bee and honey, and its many uses. In the absence of sugar, honey was an integral sweetening ingredient in Roman recipes, and references to its use in food can be found in the work of many Roman authors, including Athenaeus, Cato, and Bassus.

The art of beekeeping in ancient China has existed since time immemorial and appears to be untraceable to its origin. In the book *Golden Rules of Business Success* written by Fan Li (or Tao Zhu Gong) during the Spring and Autumn Period, some parts mention the art of beekeeping and the importance of the quality of the wooden box for beekeeping that can affect the quality of its honey.

Honey was also cultivated in ancient Mesoamerica. The Maya used honey from the stingless bee for culinary purposes, and continue to do so today. The Maya also regard the bee as sacred.Some cultures believed honey had many practical health uses. It was used as an ointment for rashes and burns, and to help soothe sore throats when no other practices were available.

Bibliography

- Apiculture in Egypt, Dr Tarek Issa Abd El-Wahab
- Apiservices - Beekeeping-Apiculture-Denmark/Danemark
- Bee Species Outnumber Mammals And Birds Combined Biology Online access date: 28/09/2009
- Beekeeping in Ancient Egypt
- Bodenheimer, F. S. (1960). *Animal and Man in Bible Lands*. Brill Archive. p. 79.
- Chandler, Philip (2007). *The Barefoot Beekeeper*. Lulu. p. 111. ISBN 978-1-4092-7114-7.
- Conradie, Beatrice & Nortjé, Bronwyn (July 2008). "SURVEY OF BEEKEEPING IN SOUTH AFRICA" (PDF). Centre for Social Science Research. Retrieved January 23, 2012.
- Crane, Eva *The World History of Beekeeping and Honey Hunting*, Routledge 1999, ISBN 0-415-92467-7, ISBN 978-0-415-92467-2, 720pp.
- Dalley, S. (2002). *Mari and Karana: Two Old Babylonian Cities* (2 ed.). Gorgias Press LLC. p. 203. ISBN 9781931956024.
- Economic aspects of beekeeping production in Croatia VETERINARSKI ARHIV 79 (4), 397-408, 2009
- Farm Commodity Programs: Honey (PDF). *nationalagl-awcenter. org*. National Honey Board. 2002. Retrieved27 March 2014.
- François Huber (1814). *Nouvelles observations sur les abeilles,*. Chez J. J. Paschoud,... et a Geneve. Retrieved 27 March 2014.
- Goodman Kurtz, Chaya (2010-06-03). "Bee-Friendly Landscaping". *Networx*.

- Gregory, Pam. "Better beekeeping in top-bar hives" (WEB ARTICLE). Bees For Development. Archived from the original on 21 March 2008. Retrieved 2008-03-12.
- Grout, Roy, A., ed. (1949). "Diseases of Adult bees". *The hive and the honey bee: a new book on beekeeping which continues the tradition of "Langstroth on the hive and the honeybee"*. Dadant and Sons. p. 607. Retrieved 21 June 2013.
- Oldest known archaeological example of beekeeping discovered in Israel (September 1st, 2008)
- Production and Value of Honey and Maple Products ; 2007
- Šljiviæ, Miljko. "beekeeping in serbia". Retrieved 27 March 2014.
- solitary bees solitary bees website, access date: 28/09/2009
- Tanguy, Marion (2010-06-23). "Can cities save our bees?". *The Guardian*.
- The Future of Bees and Honey Production in Arab Countries
- The prospects for beekeeping in the expanded EU
- Thomas Wildman, *A Treatise on the Management of Bees* (London, 1768, 2nd edn 1770).
- Traynor, Kirsten. "Ancient Cave Painting Man of Bicorp". MD Bee. Retrieved 2008-03-12.
- Walsh, Bryan (7 May 2013). "Beepocalypse Redux: Honeybees Are Still Dying — and We Still Don't Know Why". *Time Science and Space* (Time Inc). Retrieved 21 June 2013.
- Wildman, op. cit., 2nd (1770) ed., at pp. 112-115.
- Wildman, op. cit., 2nd (1770) ed., at pp. 94–95.

Index